Football Biomechanics

W0113575

Football Biomechanics explores the latest knowledge of this core discipline in sport science across all codes of the sport. Encompassing a variety of styles, including original scientific studies, syntheses of the latest research, and position statements, the text offers readers the most up-to-date and comprehensive reference of the underlying mechanics of high-level football performance.

The book is divided into five parts, covering fundamental football actions, the biomechanics of direct free kicks, footwear, biomechanical considerations in skill acquisition and training, and artificial turf. It bridges the gap between theory and practice in a variety of key areas such as:

- ball kicking mechanics (in soccer and other football codes)
- ball impact dynamics
- aerodynamics of ball flight
- special techniques (such as the 'knuckle ball shot') by world-famous players
- the efficacy and development of footwear
- biomechanical and motor performance differences between female and male soccer players
- artificial turf from an injury and a performance perspective.

Made up of contributions from leading experts from around the world, *Football Biomechanics* is a vital resource for researchers and practitioners working in all football codes, and useful applied reading for any sport science student with an interest in football.

Hiroyuki Nunome is a Professor of Biomechanics in the Faculty of Sports and Health Science, Fukuoka University, Japan. He is a former President of the Japanese Society of Science and Football and has been a member of the International Steering Group on Science and Football. He is well known for his pioneering work regarding the biomechanics of football actions.

Ewald Hennig is a Professor of Biomechanics at the Institute of Health and Biomedical Innovation, Queensland University of Technology, Australia. He is well known as patent holder for pressure distribution devices of the company NOVEL Inc., Munich, and as a world expert for footwear biomechanics relating to football.

Neal Smith is a Field Leader in Biomechanics and Research Methods at Chichester University, UK. He has recently been featured by Sky Sports in 'Ronaldo: Tested to the Limit' and by Disney XD in 'Goalmouth'.

Routledge Research in Football

Football Biomechanics

Edited by Hiroyuki Nunome,
Ewald Hennig and Neal Smith

Routledge
Taylor & Francis Group

LONDON AND NEW YORK

First published 2018 by Routledge

2 Park Square, Milton Park, Abingdon, Oxfordshire OX14 4RN
52 Vanderbilt Avenue, New York, NY 10017

Routledge is an imprint of the Taylor & Francis Group, an informa business

First issued in paperback 2019

Copyright © 2018 selection and editorial matter, Hiroyuki Nunome, Ewald Henning, and Neal Smith; individual chapters, the contributors

The right of Hiroyuki Nunome, Ewald Hennig, and Neal Smith to be identified as the authors of the editorial material, and of the authors for their individual chapters, has been asserted in accordance with sections 77 and 78 of the Copyright, Designs and Patents Act 1988.

All rights reserved. No part of this book may be reprinted or reproduced or utilised in any form or by any electronic, mechanical, or other means, now known or hereafter invented, including photocopying and recording, or in any information storage or retrieval system, without permission in writing from the publishers.

Notice:
Product or corporate names may be trademarks or registered trademarks, and are used only for identification and explanation without intent to infringe.

British Library Cataloguing-in-Publication Data
A catalogue record for this book is available from the British Library

Library of Congress Cataloging-in-Publication Data
A catalog record for this book has been requested

ISBN: 978-1-138-19512-7 (hbk)
ISBN: 978-0-367-40624-0 (pbk)

Typeset in Times New Roman
by Cenveo Publisher Services

Contents

List of contributors

Katharina Althoff works at the Institute for Sports and Movement Sciences of the University Duisburg-Essen, Germany. Her main research interests are soccer shoe development and women soccer.

Takeshi Asai is a Professor of Coaching Science and a manager of the university football team at Tsukuba University, Japan. He is well known as an expert for fluid dynamics of soccer ball.

Alexandra Atack is a Lecturer in biomechanics at St Mary's University, UK, and completed her PhD in the biomechanics of rugby place kicking. Alexandra is an active member of the International Society of Biomechanics in Sport (ISBS), British Association of Sport and Exercise Sciences (BASES) and the World Rugby Science Network.

Kevin Ball is a Senior Lecturer of the College of Sport and Exercise Science and the Institute of Sport, Exercise and Active Living (ISEAL) at Victoria University, Australia. He is currently the Australian junior AFL kicking coach, the South Sydney Rabbitohs kicking coach and consults to a number of AFL clubs. He is also on the editorial board of Sports Biomechanics, a director of the International Society of Biomechanics in Sport, an associate member of the AFL Sport Science Advisory Group and is a member of the AFL kicking skill acquisition group.

Neil Bezodis is a Senior Lecturer at Swansea University, UK, and his broad research focus is the biomechanics of elite sports performance with a focus on sprinting and rugby union. Neil is an active member of the International Society of Biomechanics in Sports (ISBS) and was recently invited to join the Editorial Board of the International Rugby Board (IRB) Rugby Science Network.

Howie J. Carson is a Researcher in motor control and sports coaching at the University of Central Lancashire, UK. As a practitioner, Howie holds 'Advanced' status as an accredited golf coach with The Professional Golfers' Association, is an accredited Sport and Exercise Scientist (specialising in skill development) through the British Association of Sport and Exercise Sciences and is a Chartered Scientist. His current research addresses the refinement of already learnt and well-established skills, skill acquisition and motor skill performance in stressful environments.

Dave Collins is a Professor of Coaching and Performance at the University of Central Lancashire, UK, and Director of Grey Matters Consultants, UK. He is a Chartered Psychologist, Registered Supervisor and Associate Fellow of the British Psychological Society, a registered practitioner with the Health Professions Council, a Chartered Scientist and a founding fellow of the Society of Martial Arts. Collins is also a Fellow of the British Association of Sport and Exercise Sciences and holds High Performance Sport Accreditation as a Sport Scientist, having worked with over 60 world and Olympic medallists plus professional performers in a variety of domains. His research interests focus on the development, evaluation and promotion of expertise in a variety of essential performance professions.

Gijs Debuyck is a Technical Youth Director in professional soccer at the Ghent University, Belgium. He was scientific collaborator at Ghent University, Belgium in a project on the biomechanical interactions between artificial turf, shoe and athlete.

Dirk De Clercq is a Professor in Biomechanics of Human Movement at the Department for Movement and Sport Sciences in Ghent University, Belgium. His research concerns biomechanics of locomotion and sports. He is head of the Laboratory for Biomechanics of Human Movement in the Sports Science Laboratory – Jacques Rogge at Ghent University, Belgium.

Eric Eils is a Professor for Performance and Training at Münster University, Germany. He is also an expert for prevention and rehabilitation of ankle injuries regarding football.

Joeri Gerlo is a staff member at the laboratory for Biomechanics and Motor Control of human movement at Ghent University, Belgium.

Sungchan Hong is a Researcher in the field of Sports Aerodynamics at Tsukuba University, Japan. In particular he has researched the effect of the panel shape of the soccer ball on its flight characteristics and fluid dynamics of soccer ball. He recently published a paper regarding the topic in *Scientific Report*.

Koichiro Inoue is a Lecturer in sport biomechanics at Yamagata University, Japan. His research interest is the dynamics of kicking motion. He won New Investigator Award in the congress of Japanese Society of Biomechanics. He is an executive board member of the Japanese Society of Science and Football.

Gareth Irwin is a Professor and Head of Biomechanics and Laboratory Director at Cardiff Metropolitan University, UK. He recently organized a FIFA research project based at the Cardiff Metropolitan University, UK. He is the president of the International Society of Biomechanics in Sports (ISBS).

Phil Kearney is a currently a Senior Lecturer in Sport and Exercise Psychology at the University of Chichester, UK, specialising in skill acquisition. His current research interests centre on technique refinement, movement pattern variability, and players'/coaches' understanding and application of core principles of skill acquisition.

Len Nokes is a Professor of Clinical Biomechanics at Cardiff University, UK, and has been involved as a FIFA consultant in coordinating a number of research programs investigating player/surface interaction.

Wolfgang Potthast is a Professor of Institute of Biomechanics and Orthopaedics at Köln Sports University, Germany. He is well known as an expert for footwear biomechanics and artificial turf. Recently, he was invited to join a FIFA research team and collaborated with FIFA to establish new FIFA standard for artificial turfs.

Stijn Rambour is a scientific collaborator at ERCAT, the artificial turf research and testing centre at the Centre for Textile Science and Engineering of Ghent University, Belgium.

Veerle Segers is a Sport Technical Director of the Flemish Basketball Federation and is a voluntary postdoctoral collaborator at Ghent University, Belgium.

Alison Sheets-Singer (PhD, Nike Inc., USA) is a Senior Researcher in the Nike Sport Research Lab in Portland, OR, USA. The goal of her research is to use biomechanical knowledge to enhance athlete performance through the creation innovative products and services. Prior to joining Nike, Inc. in 2012, Dr. Sheets-Singer was an Assistant Professor at Ohio State University, and a Postdoctoral Researcher at Stanford University. She obtained her BS from Cornell University, and PhD from University of CA, Davis in Mechanical Engineering. Sheets-Singer is a recipient of the Andrzej Komor New Investigator Award from the International Society of Biomechanics, Technical Group on Computer Simulation for her work related to optimizing gymnastics performance.

Hironari Shinkai is an Associate Professor of Sports Biomechanics at Tokyo Gakugei University, Japan. He is an executive board member of the Japanese Society of Science and Football and Editor in Chief of International online Journal *Football Science*.

Thorsten Sterzing is a Senior Consultant to the R&D Innovation Centre of Xtep Co Ltd, China, and heads their Sports Science and Engineering Laboratory. Thorsten has lectured on a wide range of the core disciplines of Sports and Human Movement Sciences at the University of Duisburg-Essen, Germany, and at Chemnitz University of Technology, Germany. He previously held the post of a Sports Science Research Manager at Li Ning Co Ltd, China, and is a patent holder in his field.

Gerda Strutzenberger is a Senior Scientist at the University of Salzburg, and was involved in a FIFA research project based at the Cardiff Metropolitan University, UK (2013–2014). Her research focuses on clinical biomechanics (sports injury, activities of daily Living) and sports performance biomechanics (sprint, cycling and football).

Ine Van Caekenberghe is a Research Associate in Biomechanics at Manchester Metropolitan University, UK, Technical Race Skills Analyst at GB Para-Swimming, UK, and voluntary postdoctoral collaborator at Ghent University, Belgium.

Stacy Winter is the Programme Director of the MSc in Applied Sport Psychology at St Mary's University, UK, and her research examines professionalism, practice, and process in applied sport psychology. Stacy is a British Psychological Society (BPS) Chartered Psychologist and a British Association of Sport and Exercise Sciences (BASES) Accredited Practitioner/Chartered Scientist.

Preface

During a football match, we sometimes witness magnificent performances by top footballers. For example, the soccer ball changes its flight path to bend around an opponent player's wall or to unpredictably move right and left, or up and down like a butterfly in flight. People in the stadium or those in front of a TV might question whether the players used some magic or if they are even free from the laws of physics. From the viewpoint of sports biomechanics the answer, however, is 'No'. Evidently, they are not inhabitants of a magic world, even though they are great practitioners of the laws of physics, whether they know it or not.

Sports biomechanics is a quantitative-based sport science discipline, playing an important role in the scientific application to the football field. In this subfield of biomechanics, the laws of physics are applied in order to gain an understanding of athletic performances through motion capture, mathematical modelling, computer simulation and other methods of measurement, using advanced technology. We are certain that the application of sports biomechanics in football, known as *Football Biomechanics*, will open your eyes to the underlying mechanics of high-level football performances.

This volume of *Football Biomechanics* includes the latest biomechanical knowledge covering football actions (Part I), direct free kicks (Part II), footwear (Part III), skill acquisition and training (Part IV), and artificial turf (Part V). Inside this volume, readers will find a unique variety of the styles of presentation including the chapters that represent adaptations of original scientific studies, and those which represent a synthesis and position statement of the current findings in that area.

Part I outlines the latest biomechanical findings about definitive actions in football. This includes four chapters covering the kicking actions of three different football codes: soccer, rugby and Australian rules, and one chapter on the action of the soccer throw-in.

Part II features the direct free kick in soccer. In this section, one chapter focuses on the aerodynamics of the ball and two chapters illustrate technical feature of special free kick (non-spinning knuckle shots), dissect the unique knuckle shot technique of two world top players, and demonstrate the application to training for achieving this technique.

Part III highlights the most important equipment used in football: footwear. One chapter describes the design, which may help to improve kicking accuracy, and the remaining three chapters provide an in-depth insight into the influence of different football shoe–surfaces and the foot–shoe interactions of the players.

Part IV explores new aspects of football biomechanics. This section includes one chapter providing a good example of evidence-based technical modification of kicking, and another chapter describing a practical framework to refine well-established kicking skills.

Part V provides useful information regarding artificial turf. This section consists of one chapter focusing on the essential property of shock absorbency, and two review chapters outlining the influence of artificial turf on injuries and performance perspectives.

Finally, we are grateful to the researchers who generously provided their latest research findings, thereby making this publication possible. We would like to thanks to the following contributors: Katharina Althoff, Takeshi Asai, Alexandra Atack, Kevin Ball, Neil Bezodis, Howie Carson, Dave Collins, Gijs Debuyck, Dirk De Clercq, Eric Eils, Joeri Gerlo, Koichiro Inoue, Gareth Irwin, Philip Kearney, Len Nokes, Wolfgang Potthast, Stijn Rambour, Veerle Segers, Alison Sheets-Singer, Hironari Shinkai, Thorsten Sterzing, Gerda Strulzenberger Sungchan Hong, Ine Van Caekenberghe, and Stacy Winter. We hope this book will stimulate further biomechanical research in the football field and encourages researchers to put theory into practice.

Part I
Football actions

Part I

Football actions

1   Does biomechanical evidence support coaching cues of kicking?

Hiroyuki Nunome

1. Introduction

As the name explains, "Football" originally referred to a variety of games that involve kicking a ball with foot to score a goal. Nowadays, we can see a diversity of footballs (association football; rugby union; rugby league; Australian rules; Gaelic football and gridiron football), all of which are known as football codes. Among these different football codes, association football (soccer) is the one that uses kicking the ball most frequently in the match. Thus, kicking is the defining action of soccer (Lees, Asai, Andersen, Nunome & Sterzing, 2010) and it is also the kicking technique that has been most widely studied from a biomechanical perspective.

Without doubt, to improve kicking performance, we want to make players kick a ball with more power. That has been the challenge for most soccer coaches. A number of practical coaching cues do exist on how to kick the ball faster, however, it seems that these coaching cues are sometimes not consistent with the scientific knowledge contained in the literature. This might confuse both researchers and coaches and thus create a gap between scientific knowledge and practical coaching cues.

While there have been many studies on the biomechanics of kicking, there are still a number of novel aspects of this skill to be explored. Recent advances in technology also helped to widen our interest to consider several new aspects of soccer kicking action, in particular actions just before and during ball impact. Our research group has made a series of novel research attempts that focused on soccer kicking. Through these studies, we obtained several unique and unexpected results, some of which support the coaching cues while others are not consistent with practical instructions.

This chapter, therefore, aims not only to give an overview of recent findings of soccer kicking biomechanics but also to shed some light on the veracity of some practical coaching instructions from a biomechanical perspective.

2. Impact phase kinematics

Leg swing kinematics at or just before ball contact during kicking are important for determining the resultant ball velocity, because there is a robust relationship that exists between the final foot velocity and the resultant ball velocity (Asami & Nolte, 1983; Barfield, 1995; Levanon & Dapena, 1998; Nunome, Lake, Georgakis, & Stergioulas, 2006). It is logical to assume that to achieve maximal ball velocity, the foot velocity should not decline before ball impact, but must be maximized until the moment the foot makes contact with the ball. However, in the literature or even in current text books of biomechanics, it has been shown that the lower leg final linear or angular velocity levels off prior to ball impact and then slightly declined towards ball impact.

Until recently, studies that have documented kicking kinematics have typically captured limb movements at rates between 100 and 400 Hz (Andersen, Dörge & Thomsen, 1999; Barfield, 1995; Dörge, Andersen, Sørensen & Simonsen, 2002; Isokawa & Lees, 1988; Lees, 1996; Lees & Nolan, 1998; Levanon & Dapena, 1998; Nunome, Asai, Ikegami & Sakurai, 2002; Rodano & Tavana, 1993; Teixeira, 1999) and then filtered the displacement data with a cut-off frequency of 6–18 Hz (Andersen *et al.*, 1999; Dörge *et al.*, 2002; Nunome *et al.*, 2002; Teixeira, 1999). Although these procedures were appropriate for illustrating the swing phase kinematics, it is unclear whether they can adequately capture and describe movement characteristics around ball impact transient. This nature of the leg swing formed our conventional understanding of the kicking leg motion. Conversely, coaches often advise players to "kick through the ball" (LA84 Foundation, 1995; National Soccer Coaches Association of America, 2012) which likely suggests that coaches encourage players to increase foot velocity until the moment of ball impact.

To address this conflict, soccer full instep kicking kinematics were captured using more advanced technology, which included a high-speed motion capture (1000 Hz) and a new filtering procedure (time-frequency filtering), which allowed changing of the cut-off frequency along the time-series (Nunome *et al.*, 2006). Key results from nine male amateur footballers are shown in Figure 1.1. The change represents shank angular velocity just before, during and just after ball impact. In the left panel shows the change using the high sampling frequency and the novel filtering, and the right panel shows the change processed using a conventional (re-sampled to 250 Hz and smoothed using a conventional filter with a low, constant cut-off frequency of 10 Hz) procedure used in the literature.

In fact, as the raw change (narrow solid line) illustrates, the shank is still angularly accelerating towards ball impact and the novel data treatment succeeded in demonstrating a more representative change without noisy oscillations on the baseline. In contrast, the conventional data treatment created a descending nature of the shank angular velocity which apparently strayed out of the raw change. This is a very important finding as this was the very pattern generally seen in the previous studies, but when a more appropriate computing procedure is employed, this pattern did not exist. From biomechanical perspective, it can be assumed that sudden deceleration caused by ball impact would produce errors in its derivative parameters in the last few frames before ball impact and this would be amplified if smoothing was attempted through impact. This was a primary source of misleading ball kicking biomechanics. However, most ball kicking studies have

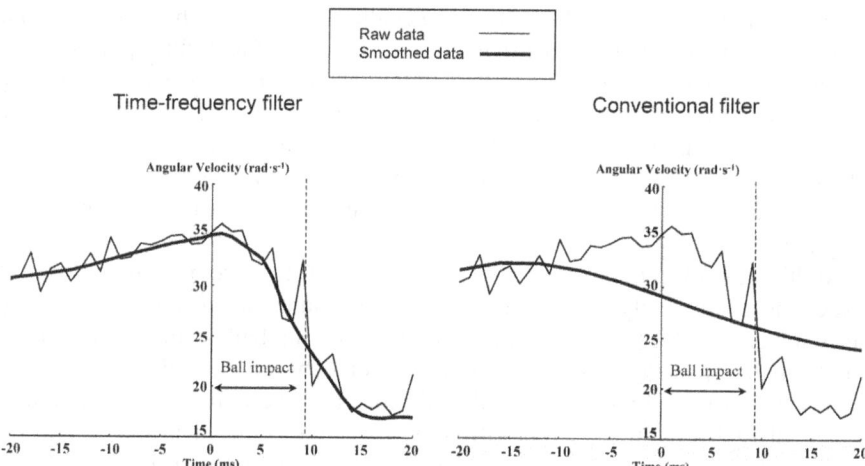

Figure 1.1 Comparison of shank angular velocity changes just before, during and after ball impact from two different filtering and sampling procedures. In the left panel, high-speed sampled data (1000 Hz) smoothed by a time-frequency filter is shown. In the right panel, raw data was re-sampled (250 Hz) and then smoothed by a conventional filter at 10 Hz cut-off frequency is shown (modified from Nunome *et al.* (2006) in *Journal of Sports Sciences*)

failed to acknowledge this type of potential error, and it is often unclear how displacement coordinates were smoothed (Isokawa & Lees, 1988; Rodano & Tavana, 1993; Lees, 1996; Lees & Nolan, 1998).

This work was the first to provide evidence that strongly supports the above practical advice of kicking by revealing a 'more representative' shank motion around ball impact. The coaching perspective of "kicking through the ball" would indeed be a cue that has direct application to actual kicking performance.

3. Ball impact dynamics

There are several instructions for kicking the ball with power, in which the importance of a good follow through has been frequently stressed (O'Challagan, 2010). Barfield (2000) suggested that one primary objective of the follow through is to keep the foot in touch with the ball as long as possible. Coaches, players and even researchers believe that skilled players are capable of pushing the ball longer than less skilled players, thereby increasing the resultant ball velocity. This coaching cue has a sound theoretical underpinning because the final momentum of the ball is determined by the amount of impulse applied to the ball (impulse-momentum theorem). If players can apply more impulse to the ball by lengthening ball contact time, the ball will gain more momentum, namely a "faster" resultant ball velocity to be achieved.

Ball impact transient has been treated as something of a "black box" and thus to explore the inside of the black box is a challenge for researchers. There is one study which treated ball impact dynamics using a very fast (4800 fps) sampling frequency (Tsaousidis and Zatsiorsky, 1996). Their study, however, had limitations within the mechanical models of the foot and the ball that would prevent a realistic examination of typical kicks in soccer. Thus, our research group made an attempt to describe foot-ball dynamic interaction during ball impact more precisely by quantifying the ball deformation and the displacement of its centre of gravity.

Multi-angle ultra-high speed video records (5000 fps) were obtained from 11 experienced university male soccer players. A mathematical model was developed to compute the centre of gravity of the deforming ball. In this mathematical model, ball impact was assumed to be a collision between the ball and a flat plane which was perpendicular to the ball trajectory and the ball was modelled as a spherical shell in which the mass (0.43 kg) was uniformly distributed onto the surface. The non-deformed part of the ball was constructed using consecutive hollow circular discs, and the centre of the gravity of the non-deformed part was computed as a function of ball deformation. Also, for the deformed part, its centre of gravity was assumed to locate the centre of the flat contact plane. From this model, the trajectory of the centre of gravity of a deforming ball during ball impact was obtained. Further detailed computing procedure is explained in Shinkai, Nunome, Isokawa and Ikegami (2009).

Figure 1.2 shows the time-series change of foot (CG) velocity, ball (CG) velocity and ball deformation during ball impact. These changes suggest intriguing aspects of ball impact dynamics and provided new insight for practical instructions. Although the ball impact phase has quite a short duration (approximately 9 ms) the ball impact can be divided into four phases. Until the end of phase II, the foot velocity always exceeds that of the ball, suggesting the foot accelerates the ball by imparting its linear momentum to the ball. Then, the ball CG velocity begins to exceed that of the foot after the ball was maximally deformed and the ball decompression begins in phase III. In this phase, it can be presumed that the ball recoil occurs on the foot and is probably the main contributor to increasing ball velocity to approximately (95%) launch velocity. At the beginning of phase IV, although it seems that the foot still has contact with the ball, there is no dynamic interaction between the two bodies. Thus, it should be interpreted that the effective duration, in which players can positively contribute to increase ball velocity, is approximately less than half of the contact time.

As mentioned before, practical advice such as "push the ball as long as possible" sounds logically consistent to achieve faster ball velocity. However, our findings indicated that players in all skill levels are incapable of controlling the ball recoiling on the foot which happens after the ball has maximally deformed.

To confirm the relationship between ball contact time and resultant ball velocity more precisely, we attempted to mimic the soccer ball impact under controlled

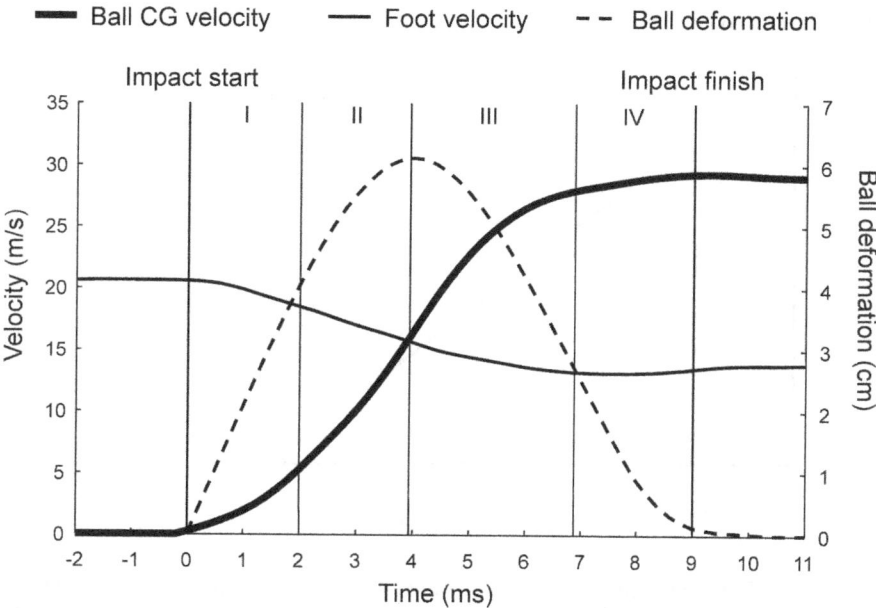

Figure 1.2 Changes of foot (CG) velocity, ball (CG) velocity and ball deformation during ball impact (reused with permission from Shinkai, H., Nunome, H., Isokawa, M. & Ikegami, Y., "Ball impact dynamics of instep soccer kicking" in *Medicine and Science in Sports Exercise*, published on 1st January, 2009, Wolters Kluwer Health, Inc.)

conditions. In Iga, Nunome and Ikegami's (2013) study, a regular soccer ball was directly fired from a launching machine at a force platform fixed vertically on a pedestal with five different velocities. The sagittal motion of the ball was captured at an ultrahigh sampling speed (5000 fps) and from the combination of ball impact force and video footage, ball contact time and resultant ball velocity was computed.

As expected, the ball contact time systematically decreased as the ball velocity increased (Figure 1.3), which is conflicting with the practical advice to "push the ball as long as possible" to increase time on the ball. This result confirmed that lengthening contact time during ball impact will not lead faster resultant ball velocity in soccer ball kicking.

The nature of the ball impact dynamics have clarified a gap between what players try to do and phenomena actually happening on the foot during ball impact. However, this finding needs to be carefully interpreted from two sides: practical and scientific perspectives. While from a scientific point of view, increasing contact time is useless for improved performance, the coaching cue "push the ball as long as possible" might still be an easy, clear target towards refining players' kicking technique, in particular for novice players.

Contact time (ms)

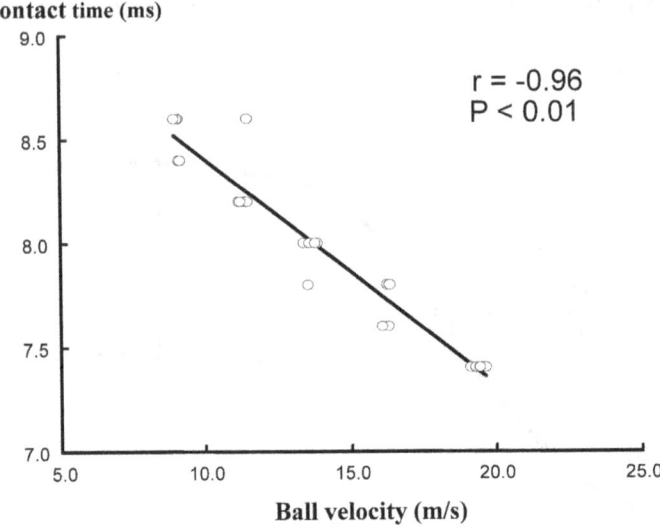

Figure 1.3 The relationship between ball velocity and contact time (modified from Iga *et al.* (2013) in *Proceedings of the 31st Conference of the International Society of Biomechanics in Sports*)

4. Side-foot kicking

Among many forms of soccer kicking, the side-foot kick is the most frequently used technique when a short and precise pass or shot is required; for example, the side-foot kick was used to score 16 of 17 goals from penalty kicks in the World Cup 1998 (Grant, Reilly, Williams & Borrie, 1998).

Side-foot kicking is a fundamental technique. However, beginners often have difficulty in learning side-foot kicking technique due to the unusual leg motion: to swing the medial side of the foot forward. Thus, a number of practical instructions exist to introduce novice players to the side-foot kicking technique. For example, the side-foot kick is often called the "push pass", and coaches often advise novice players to "push the heel towards the target".

Soccer kicking has been widely studied, however, most studies were performed on the quasi-planar aspects of the instep kick or similar maximal soccer kicks (Roberts, Zernicke, Youm & Huang, 1974; Asami & Nolte, 1983; Robertson & Mosher, 1985; Luhtanen, 1988; Putnam, 1991). In order to understand how the leg swing is produced in side-foot kicking and how authentic the common practical instructions are for learning side-foot kicking, we tried to identify the three-dimensional kinetics of five experienced male high school players performing side-foot kicks (Nunome, *et al.*, 2002).

A remarkable difference was observed between the side-foot and instep kicks for the hip external rotation torque and angular velocity (Figure 1.4). The hip

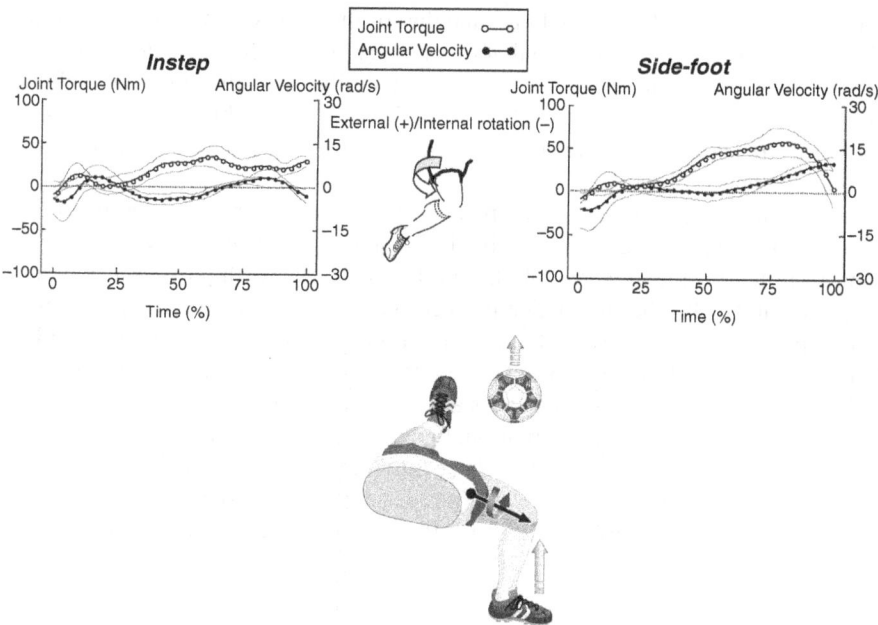

Figure 1.4 Changes in hip external /internal joint torque and angular velocity of the side-foot (right column) and instep (left column) kicks (top panels), and top view of hypothetical mechanics of the side-foot kick (bottom panel) (modified from Nunome *et al.* (2002) in *Medicine and Science in Sports Exercise*)

external rotation torque was dominant only for the side-foot kick and its magnitude was significantly larger than that of the instep kick. Also, the hip external rotation angular velocity of the side-foot kick increased rapidly during the final phase of leg swing, whereas that of the instep kick was distinctively small. The magnitude of hip external rotation angular velocity of the side-foot kick was also significantly larger than that of the instep kick. It is reasonable to consider that this marked difference is closely associated with a coaching cue "push the heel towards the target". It is evident that knee joint planar motion is not available to swing the medial side foot forward because the knee joint does not bend in a sideways direction. One possible solution is to twist the hip joint externally.

As shown in Figure 1.4 (bottom panel), we observed a typical leg configuration in the final phase of side-foot kicking among high school players: the plane formed by thigh-shank segment is pointing outward while the knee is not fully extended. In this configuration, the players can utilize the hip external rotation motion to directly increase the forward velocity of the medial side of the foot. In conjunction with these findings, coaching cues regarding side-foot kicking should emphasize the hip external rotation.

However, the side-foot kick of an international-level player demonstrated an apparently different style of kicking than that of the high school players, and his kicking style was quite similar to that of a novice player who just started playing soccer. Interestingly, it can be seen that a vast similarity of the side-foot kicking motions between the novice and the international players while the high school player demonstrated a unique style of kicking. This discrepancy is underpinned by the kinetics.

As shown in Figure 1.5 (right top panel), the hip external rotation torque of the international player declined rapidly during the final phase of leg swing in contrast to that of the high school players. Interestingly, the hip external rotation torque exhibited by the international-level player was quite similar to that of a novice player who had just started learning how to kick the ball with their side-foot (Figure 1.5 right bottom panel). We observed large differences between the kicking kinetics of high school players and the international player, in contrast to the similarities between the international player and the novice player.

Caution is need in interpreting these unique and unexpected results seen between the three different skill levels (novice, skilled and exceptionally skilled). One view is that the coaching cue "push the heel towards the target" would be a

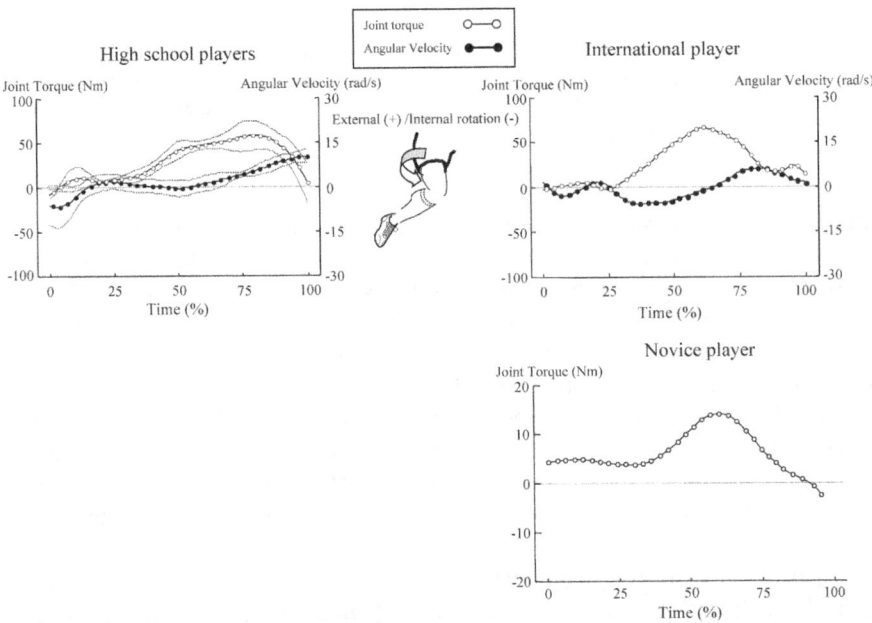

Figure 1.5 Changes in hip external /internal joint torque and angular velocity of side-foot kick. Left panel shows the data of high school players, right top panel shows that of an international player and right bottom panel shows that of a novice player (joint torque data only) (modified from Nunome *et al.* (2002) in *Medicine and Science in Sports Exercise*)

good aid to achieve the side-foot kicking style generally seen for skilled players. However, we should recognize the possibility that novice players who already have comparable side-foot kicking mechanics with that of exceptionally skilled players may not necessarily need to receive these kinds of coaching cues. Perhaps, coaches should consider two factors; 1) help players to achieve the fundamentals of side-foot kicking, and 2) avoid stereotyping or defining a player's side-foot kicking style through excessive coaching cues.

5. Conclusion

The validity of three practical coaching instructions has been investigated from a biomechanical perspective. The coaching cue "kick through the ball" obtained solid support while "push the ball as long as possible" demonstrated a conflict with the phenomena that actually happened during ball impact. Also the coaching cue that encourages players to learn a side-foot kick showed consistency for the kicking mechanism of skilled players but was not consistent with that of a player in exceptional skill level and of a novice player.

References

Andersen, T. B., Dörge, H. C. & Thomsen, F. I. (1999). Collisions in soccer kicking. *Sports Engineering*, 2, 121–125.

Asami, T., & Nolte, V. (1983). Analysis of powerful ball kicking. In H. Matsui, & K. Kobayashi (Eds.), *Biomechanics VIII-B* (pp. 695–700), Champaign, IL: Human Kinetics.

Barfield, W. R. (1995). Effects of selected kinamatic and kinetic variables on instep kicking with dominant and nondominant limbs. *Journal of Human Movement*, 29, 251–272.

Barfield, W. R. (2000). Biomechanics of kicking. In W. E. Garret, & D. T. Kirkendall (Eds.), *Exercise and Sports Science* (pp. 551–562), Philadelphia: Lippincott Williams & Wilkins.

Dörge, H.C., Andersen, T.B., Sørensen H. & Simonsen, E.B. (2002). Biomechanical differences in soccer kicking with the preferred and the non-preferred leg. *Journal of Sports Sciences*, 20, 293–299.

Grant, A., Reilly, T., Williams, M. & Borrie, A. (1998) Analysis of the goal scored in the 1998 world cup. *Insight: The FA Coaches Association Journal*, 2, 18–20.

Iga, T., Nunome, H. & Ikegami, Y. (2013). Basic mechanical analysis of soccer ball impact. In *Proceedings of the 31st Conference of the International Society of Biomechanics in Sports*, Taipei: National Taiwan Normal University. Available from: https://ojs.ub.uni-konstanz.de/cpa/article/view/5637/5130.

Isokawa, M. & Lees, A. (1988). A biomechanical analysis of the instep kick motion in soccer. In *Science and football* (pp. 449–455), London: E & FN Spon.

LA84 Foundation. (1995). Soccer coaching manual. Available from: http://library.la84. org/3ce/CoachingManuals /LA84soccer.pdf.

Lees, A. (1996). Biomechanics applied to soccer skills. In T. Reilly, A. Lees, K. Davids & W. J. Murphy (Eds.), *Science and Football* (pp. 123–134), London: E & FN Spon.

Lees, A. & Nolan, L. (1998). The biomechanics of soccer: A review. *Journal of Sports Sciences*, 16, 211–234.

Lees, A., Asai, T, Andersen, T.B., Nunome, H. & Sterzing, T. (2010). The biomechanics of kicking in soccer: A review. *Journal of Sports Sciences*, 28, 805–817.

Levanon, J. & Dapena, J. (1998). Comparison of the kinematics of the full-instep and pass kicks in soccer. *Medicine and Science in Sports Exercise*, 30, 917–927.

Luhtanen, P. (1988). Kinematics and kinetics of maximal instep kicking in junior soccer players. In: T. Reilly, A. Lees, K. Davids & W. J. Murphy (eds.), *Science and Football* (pp. 441–448). London: E & FN Spon.

National Soccer Coaches Association of America. (2012) Section 4: Basic Skills – Receiving, Passing and Shooting. Available from: http://www.nscaa.com/education/resources/fundamentals/basic-skills-receiving-passing-shooting.

Nunome, H., Asai, T., Ikegami, Y., & Sakurai, S. (2002). Three dimensional kinetic analysis of side-foot and instep kicks. *Medicine and Science in Sports and Exercise*, 34, 2028–2036.

Nunome, H., Lake, M., Georgakis, A. & Stergioulas, L.K. (2006). Impact phase kinematics of the instep kick in soccer. *Journal of Sports Sciences*, 24, 11–22.

Nunome, H., Ball, K. & Shinkai, H. (2014). Myth and fact of ball impact dynamics in football codes. *Footwear Science*, 6, 105–118.

O'Challagan, J. (2010). How to kick a soccer ball. THE COMPLETE Soccer Guide. Available from: http://www.completesoccerguide.com/how-to-kick-a-soccer-ball/.

Putnam, C.A. (1991). A segment interaction analysis of proximal-to-distal sequential segment motion patterns. *Medicine and Science in Sports and Exercise*, 23, 130–144.

Roberts, E.M., Zernicke, R.F., Youm, Y. & Huang, T.C. (1974). Kinetic parameters of kicking. In: R. C. Nelson & C. A. Morehouse (Eds.), *Biomechanics IV* (pp. 157–162). Baltimore: University Park.

Robertson, D. G. E. & Mosher, P. E. (1985). Work and power of the leg muscles in soccer kicking. In: D. A. Winter, R. W. Norman, R. P. Wells, K. C. Hays & A. E. Patla (Eds.), *Biomechanics IX-B* (pp. 533–538). Champaign, IL: Human Kinetics.

Rodano, R. & Tavana, R. (1993). Three dimensional analysis of the instep kick in professional soccer players. In T. Reilly, J. Clarys, & A. Stibbe (Eds.), *Science and Football II* (pp. 357–361), London: E & FN Spon.

Shinkai, H., Nunome, H., Isokawa, M. & Ikegami, Y. (2009). Ball impact dynamics of instep soccer kicking. *Medicine and Science in Sports and Exercise,* 41, 889–897.

Teixeira, L.A. (1999). Kinematics of kicking as a function of different sources of constraint on accuracy. *Perceptual and Motor Skills*, 88, 785–789.

Tsaousidis, N. & Zatsiorsky, V. (1996). Two types of ball–effector interaction and their relative contribution to soccer kicking. *Human Movement Science*, 15, 861–876.

2 The support leg motion as a great contributor to football kicking

Koichiro Inoue

1. Introduction

Instep kicking is well known as one of the most fundamental and important techniques in association football. Players commonly use this technique when kicking the ball with power. Previous studies about the biomechanics of instep kicking have mainly considered the kicking leg action (Apriantono *et al.*, 2006; Barfield *et al.*, 2002; Dörge *et al.*, 2002; Levanon & Dapena, 1998; Luhtanen, 1988; Nunome *et al.*, 2002; Nunome *et al.*, 2006a; Rodano & Tavana, 1998). As a strong, positive correlation exists between the velocity of kicking foot immediately before ball impact and the resultant ball velocity (Andersen & Dörge, 2011; Asami & Nolte, 1983; Isokawa & Lees, 1988; Levanon & Dapena, 1998; Nunome *et al.*, 2006b), investigators have focused on mechanisms of how to achieve faster foot velocity during the swing motion of the kicking leg towards ball impact.

Conversely, well-informed coaches pay attention to the support leg motion as well as the kicking leg so as to improve instep kicking technique. They often advise their players towards appropriate motions of the support leg, such as the position of a pivot foot and the configuration of support leg joints. To date, the support leg has gathered little interest in the research literature (Lees *et al.*, 2009; Kellis *et al.*, 2004). However, the mechanics of the support leg motion have not been revealed three-dimensionally while the kicking leg motion has been characterised by detailed segmental and joint motions in multiple planes (Kellis & Katis, 2007). From the point of view of dynamics, the support leg is the only body part that receives external force (ground reaction force) reaching more than twice the kicker's body weight (Katis & Kellis, 2010; Kellis *et al.*, 2004). However, no information has been provided about how the support leg adapts to such a large force.

It is widely accepted that a matured instep kick motion includes proximal-to-distal sequential motion called the "kinetic chain" or "whip-like motion". Horizontal pelvic rotation existing during the instep kicking (Levanon & Dapena, 1998; Nunome *et al.*, 2002), can be considered to be an incipient motion prior to the sequential swing motion of the kicking leg. The support leg motion would initiate the pelvis rotation, thereby contributing to the sequential motion kicking leg, yet the mechanical interaction between the support leg and the pelvis segment has never been clarified.

This chapter discusses the role of the support leg during the instep kicking in football, based on our recent work (Inoue *et al.*, 2014). We aim to provide information which will describe the kinetic aspects of the support leg motion in detail as well as its essential functions during kicking. The aim of this chapter, therefore, was to examine the mechanics of the support leg motion during football maximal instep kicking, using three-dimensional motion analysis.

2. Methods

2.1 Subjects

Fifteen male experienced collegiate soccer players (age = 20.9 ± 0.5 years, height = 170.2 ± 4.9 cm, body mass = 67.9 ± 7.6 kg, soccer experience = 14.1 ± 1.9 years with a minimum of 10 years; Mean ± SD) from a team in a top regional collegiate league, volunteered to participate in the present study. All participants preferred to kick the ball with their right leg. The experiment protocol was approved by the human research committee. Informed written consent was obtained from each participant before the experiment.

2.2 Experimental procedure

Kicking motions were captured using a motion capture system (Vicon T20; Vicon Motion Systems, Oxford, UK) at 500 Hz. Ground reaction forces underneath the support leg were also recorded simultaneously by a bare force platform (Type 9281E; Kistler Instruments, Winterthur, Switzerland) embedded in the floor. Three-dimensional coordinates were expressed as a right-handed orthogonal reference frame fixed on the floor (Z was vertical and pointed upward; Y was horizontal and pointed in the direction of the kicking direction, while X was perpendicular to Y and Z).

A FIFA-approved size 5 football (Jabulani Lushiada; ADIDAS Japan K.K., Tokyo, Japan) was used and its inflation was controlled at 900 hPa. Participants wore the same type of rubber outsole soccer shoes, socks, a compressive shirt and shorts, although differing in size. Passive reflection markers were fixed firmly by double-sided tape onto 21 body landmarks.

After warm-up, the participants were instructed to perform maximum effort instep kicks of a stationary ball using their preferred leg (right leg). Kicks were performed towards a target 6 m away. The approach run up was standardised to three steps. Two successful shots for each participant (having a good foot–ball impact, the support leg foot position on the force platform and getting the centre region of the goal) were sampled for subsequent analysis.

2.3 Calculation of parameters

Hip joint centres were defined following the procedure described by Davis *et al.* (1991) who developed a function to estimate hip joint centres using radiographic measurement data of pelvis and leg length data. The knee joint centres were defined as the midpoint between the lateral and medial epicondyle of the femur. The ankle joint centres were defined as the midpoint between the lateral and medial malleolus. The support leg was modelled as a four-link segment composed of the foot, the shank, the thigh and the pelvis segment. Segment and joint reference frames which fixed on each segment or joint of the lower legs including pelvis were defined according to Nunome *et al.* (2002).

The joint angular velocity vectors were calculated the method based on the previous studies (Feltner & Nelson, 1996; Sprigings *et al.*, 1994). Segment mass, the centre of mass location, and inertial values were derived from the data of young living Japanese athletes (Ae *et al.*, 1992) which was considered to be the most appropriate for the participants in the present study. The joint torque vectors of each joint were calculated in accordance with inverse dynamics (Winter, 2009). The joint angular velocity vectors and joint torque vectors of each joint were separated into anatomically relevant components according to the joint reference frames. The support leg was presented as having seven degrees of freedom: dorsal flexion (+)/ plantarflexion (–), inversion (+)/ eversion (–), adduction (+)/ abduction (–) at the ankle joint; extension (+)/ flexion (–) at the knee joint; adduction (+)/ abduction (–), flexion (+)/ extension (–), external (+)/ internal (–) rotations at the hip joint.

In order to describe the influence of the support leg on the pelvis rotation motion, we computed the torques acting from the support leg: the reaction hip joint torque (RJT) and the torque due to the hip joint reaction force (TJRF). RJT was defined as a reversal torque of the hip joint torque. TJRF was defined as an interaction torque, which computed by the cross product of the moment arm (the vector from the middle point of both hip joint centres to the support leg hip joint centre) and the hip joint reaction force acting on pelvis segment (Figure 2.1). Lastly, those torques (RJT and TJRF) were calculated as a component of the pelvis vertical axis, which defined a vector to the pelvic plane formed by both hip joint centres and the

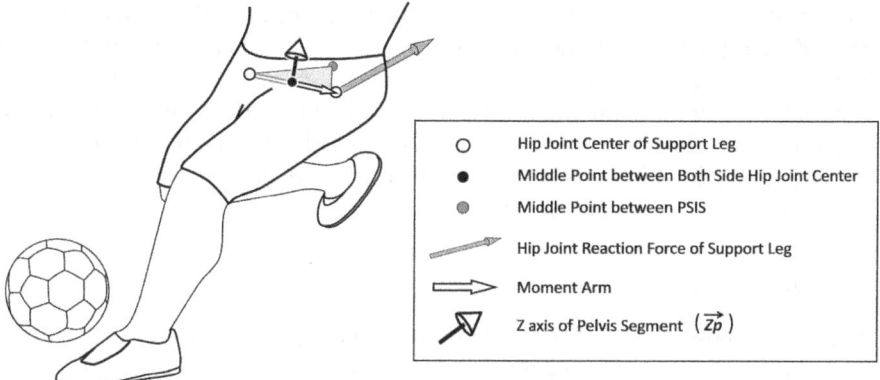

Figure 2.1 Explanation of the parameters for calculating the torque due to hip joint reaction force. Modified from Inoue *et al.* (2014) in *Journal of Sports Sciences*

midpoint of anterior posterior iliac spine (i.e. a yaw axis of the pelvis segment). The angular velocity of the pelvis segment was also computed as the rate of rotation about the vertical axis. Counter-clockwise rotation from the overhead view corresponded to the positive direction (+) for the angular velocity and torques.

All the torque, angular velocity and ground reaction force data were calculated from the raw three-dimensional coordinate data and/or the raw force platform data. The joint torque data were then smoothed using a fourth-order Butterworth low-pass filter with a cut-off of frequency of 25 Hz. From observation it was noted that angular velocity and ground reaction force data contained minimal amounts of noise. For this reason, we judged that it was not necessary to smooth angular velocity and the ground reaction force data.

According to the definition described by Lees *et al.* (2009), the touch-down of the support leg was defined as occurring when the vertical ground reaction force had exceeded 20 N. The period from the touch-down of the support leg to ball impact was normalised to 100%. To illustrate the data before the touch-down, the analysed portion was expanded to −50% using the same scaling factor for normalization. For instance, if the total time from the touch-down to the ball impact is 0.10 s, the start point of the data will be 0.05 s before the touch-down. The period from −50% to 0% (before the touch-down) and that from 0% to 100% (from the touch-down to the ball impact) were termed flight phase and support phase, respectively.

In the present study, calculation for all parameters was performed on each trial (a total of 30 kicks), and then the values were used to compute discrete mean and ensemble average curves.

3. Results and discussion

3.1 Support leg kinetics

Figure 2.2, 2.3 and 2.4 show the average value of the joint angular velocities and torques of each joint on the support leg. The joint torques calculated for the present study represent the muscle torque that is generated by the muscles located around each joint. Hence, the phase where the angular velocity and torque have the same sign, means that concentric contraction exerts the positive power on the joint. In contrast, when these values have conflicting signs, the joint generate the negative power by eccentric contraction. Additionally, the angular velocity with accompanying little joint torque represents the passive motion occurring from an external force other than muscle force.

3.1.1 Ankle joint

Figure 2.2 shows the average value of the joint angular velocities and torques of the ankle joint on the support leg. At the earlier part of the support phase (from 0% to 30%), the ankle experienced a significant and rapid multi-axial motion, which did not accompany the action of the joint torques. After this, those actions are reduced toward ball impact, while increasing plantarflexion torque.

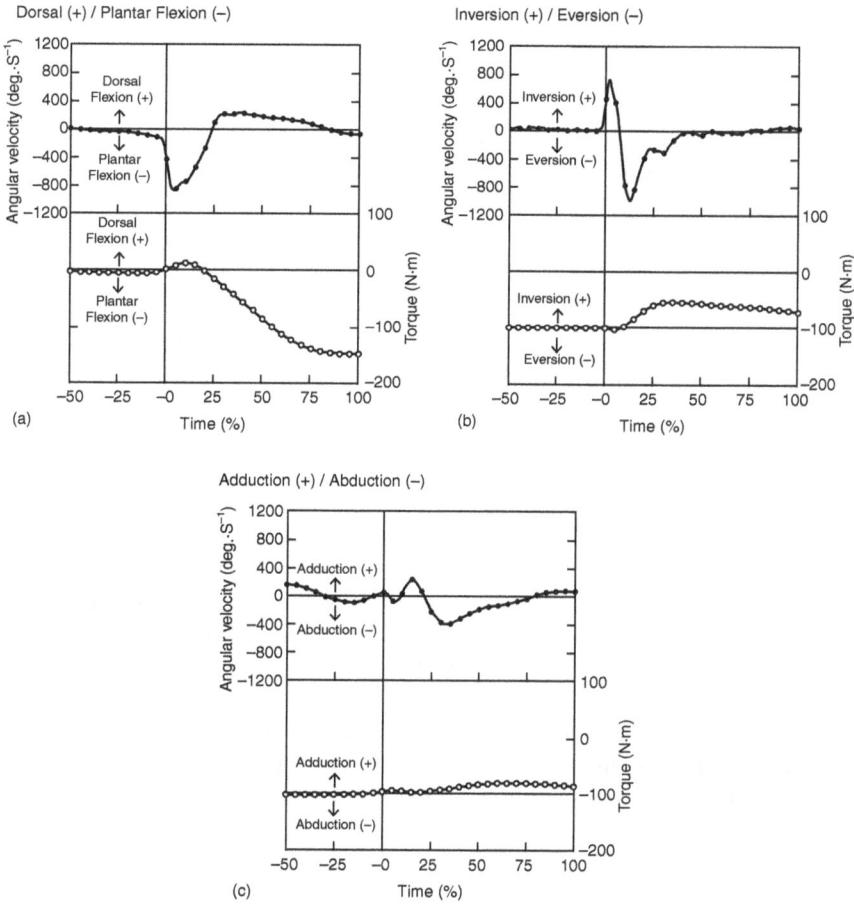

Figure 2.2 The average value of the joint angular velocities (top) and torques (bottom) at the ankle joint of the support leg: dorsal flexion/ plantarflexion (a), inversion/ eversion (b) and adduction/ abduction (c). Modified from Inoue *et al.* (2014) in *Journal of Sports Sciences*

From the approximate instant of the touch-down, the ankle was rapidly forced into plantarflexion and inversion. Immediately thereafter, the ankle was forced into opposite directions: dorsal flexion and eversion. Meanwhile, the ankle joint did not generate the remarkable joint torques for leading those rapid motions. Previous studies reported the support leg received large ground reaction force (Katis & Kellis, 2010; Kellis *et al.*, 2004); and a similar result was confirmed in the present study as well. It is logical to assume that these rapid motions of the ankle joint were triggered by the impact peak of the ground reaction force typically seen just after touch-down. In other words, the three-dimensional motion observed on the ankle joint was generated passively, not by muscle actions but

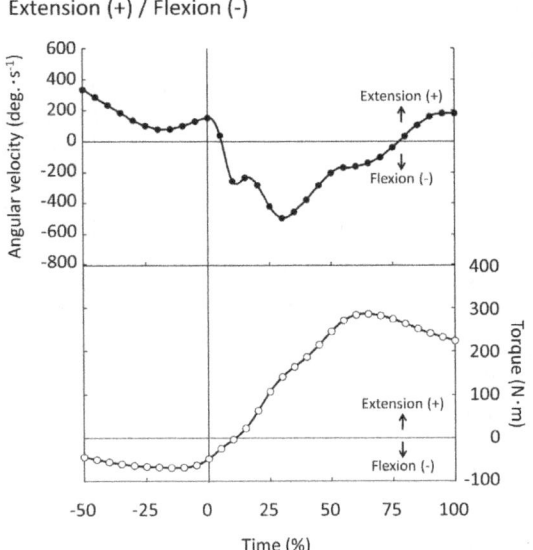

Figure 2.3 The average value of the joint angular velocity (top) and torque (bottom) at the knee joint of the support leg: extension/ flexion. Modified from Inoue *et al.* (2014) in *Journal of Sports Sciences*

rather the ground reaction force. As the ankle joint could not be actively controlled during the support phase, it might be important during the mature instep kick to adjust the foot segment posture before touch-down.

3.1.2 Knee joint

Figure 2.3 shows the average value of the joint angular velocity and torque of the knee joint on the support leg. The joint torque counteracted the joint motion for the most of the movement, except for the phase immediately before ball impact (from 75% to 100%). During the flight phase, the flexion torque was dominant while the support leg knee was extending toward touch-down. Soon after the touch-down, the knee joint torque developed a large extension torque while the knee was flexing. Immediately before ball impact, the extension torque remained and the knee came to extend again.

During most of the support phase, the knee joint exhibited negative power composed of eccentric contraction of the knee extensors. It can be interpreted that the muscle activation mainly contributes to absorb the shock of touch-down and to resist the large external force such as the ground reaction force in order to stabilise the body. It is suggested that a certain level of knee extensor strength is needed for conducting a stable instep kicking motion.

Immediately before ball impact, the knee extention motions came to be associated with the knee extension joint torque, thereby producing a distinctive positive power.

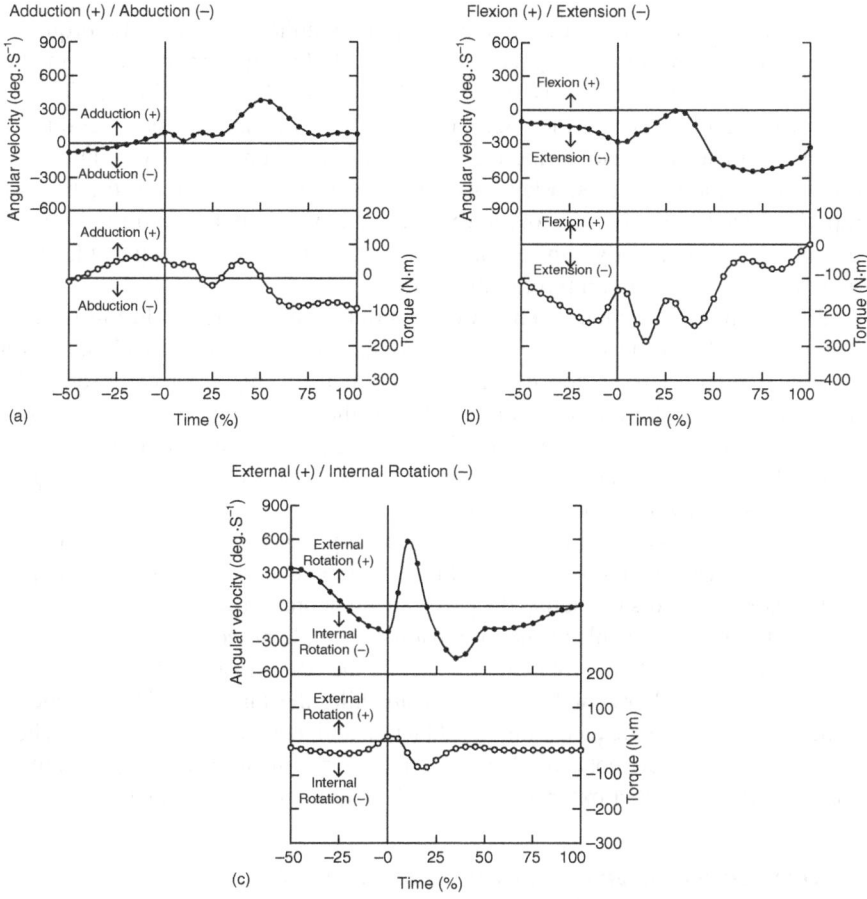

Figure 2.4 The average value of the joint angular velocities (top) and torques (bottom) at the hip joint of the support leg: adduction/ abduction (a), extension/ flexion (b) and external rotation/ internal rotation (c). Modified from Inoue *et al.* (2014) in *Journal of Sports Sciences*

Nunome and Ikegami (2005) demonstrated that linear upward acceleration of the hip joint on the kicking leg side induces a motion-dependent interaction torque for accelerating the kicking leg swing at the final phase of kicking. In the present study, the positive power due to the knee extension torque was seen on the support leg just before the ball impact, suggesting the knee extensors of the support leg were exhibiting concentric contraction during this phase. Those muscle activations most likely serve to lift the body and contribute to produce the linear upward acceleration of the hip joint on the kicking leg side. Practically, in order to achieve a fluent action of the motion-dependent interaction torque acting on the kicking leg, lifting the whole body upward by active knee extension of the support leg would be an effective action, in particular during the final phase of kicking.

3.1.3 Hip joint

Figure 2.4 shows the average value of the joint angular velocities and torques of the hip joint on the support leg. All through the flight and support phases, the hip constantly exhibited an extension torque. Meanwhile, hip extension motion was almost maintained during kicking. Although there was a noteworthy adduction angular velocity increase during the middle of the support phase (around 50%), the joint torque was not substantially large. From the touch-down, the hip was rapidly forced into external rotation while a slight internal rotation torque was exhibited. Subsequently, an apparent internal hip rotation appeared, while generating little internal hip rotation torque.

The positive power had been almost continuously generated during kicking. This indicated that the support leg made contact with the ground with active hip extension referred from concentric contraction using hip joint extensors. This action resembles the hip joint motion during the stance phase of running. It is likely that this active hip extension with positive power works to move the body forward in the same manner as a running motion; and contributes to keeping the approach velocity after the touch-down.

As the foot segment of the support leg is mostly fixed on the ground throughout the support phase, it can be considered that the hip adduction and internal rotation motion are key factors in rotating the pelvis segment onto a transverse plane. For this reason, it seems reasonable to suppose that the multiple hip joint motions including the adduction and internal rotation observed in the present study, should be associated to the pelvis horizontal rotation. Meanwhile, the hip joint did not generate noticeable torques associated with the adduction and internal rotation motion. Those motions, therefore, are not caused by the muscular system around the hip joint of the support leg but by other external factors such as ground reaction forces.

3.2 Interaction between the support leg and the pelvis

Figure 2.5 shows the average value of the pelvis rotation angular velocity and the two kind of torques acting from the support leg to pelvis segment (RJT and TJRF). There was a clear onset of pelvis counter-clockwise rotation (overhead view) in the point in time of touch-down. After reaching its peak magnitude during the mid support phase, the angular velocity decreased during the latter half of the support phase. TJRF rapidly increased after touch-down and acted in the counter-clockwise direction during the entire support phase. In contrast, RJT began to act on the clockwise direction from the mid support phase (around 50%).

Pelvis horizontal rotation is typically seen in the matured instep kicking style and its importance has been described in previous studies (Roberts & Metcalf, 1968; Wickstrom, 1975). It has been characterised that the rotation of the pelvis precedes a proximal-to-distal sequence motion of the kicking leg. While it can be seen that the pelvis angular velocity did not exhibit a marked change before the touch-down, a sudden transition of the pelvis angular velocity was observed from the instant of touch-down of the support leg after which angular velocity

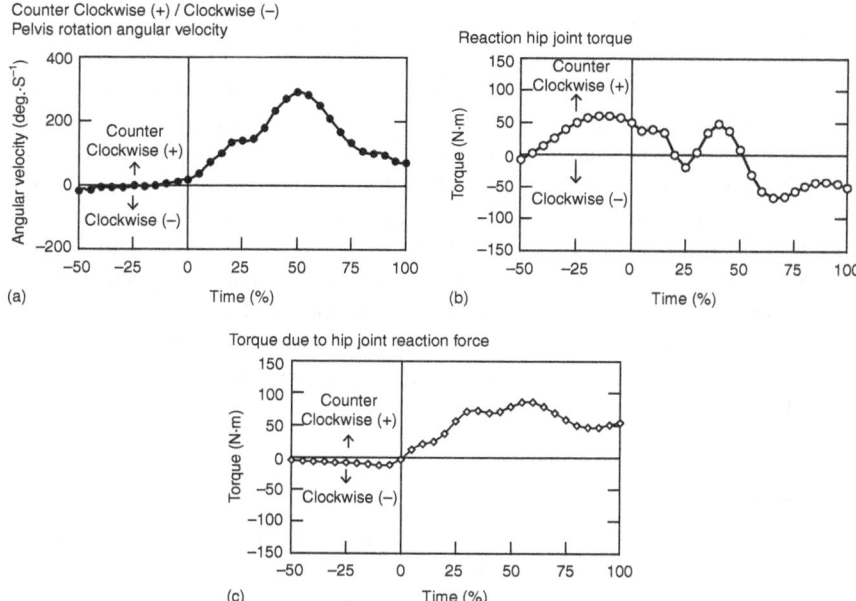

Figure 2.5 The average value of the pelvis rotation angular velocity (a) and the two torque acting on the pelvis from the support leg: RJT, namely the reaction joint torque (b) and TJRF, namely the torque due to the hip joint reaction force (c). Modified from Inoue *et al.* (2014) in *Journal of Sports Sciences*

continued to increase until the middle of the support phase. This motion coincided with a clear increase of TJRF, while there were no appreciable changes for RJT. As the touch-down of the support leg seemed to trigger the rapid pelvis rotation, the joint reaction force of support leg side due to the ground reaction force was the most dominant factor to induce this counter-clockwise pelvis rotation. This result suggests that instruction in the support leg motions may lead to an improvement of pelvis rotation motion.

After the mid support phase, the angular velocity of the pelvis decreased while TJRF still maintained a distinctive positive value. From the onset of the decrease of the pelvis angular velocity, RJT began to generate a clockwise torque. Nunome *et al.* (2006a) indicated that the deceleration of the kicking leg thigh typically seen in the proximal-to-distal sequences was initiated by the kicking leg knee joint torque reversely acting on the thigh. The result of the present study falls in line with the result of the previous study for the cause of kicking leg thigh deceleration. The result of the present study indicated that RJT initiated the deceleration of the pelvis rotation, thereby forming the apparent proximal-to-distal sequential motion between the pelvis and the kicking leg. Thus, the muscles around the hip joint of the support leg have a function to inhibit the pelvis rotation during the latter half of support phase.

4. Conclusions

It can be concluded that: 1) drastic ankle plantar flexion, ankle inversion/ eversion motions appear just after the support leg touch-down were most likely triggered by the impact peak of the ground reaction force, 2) knee joint extension torque produced with knee flexion motion has a role in absorbing the shock of the landing, 3) knee extension motions immediately before ball impact were accompanied by those extension joint torques and would contribute to acceleration of the swing of kicking leg, and 4) the torque due to the hip reaction force from the support leg attributing to the ground reaction force was the main factor in producing counter-clockwise rotation of the pelvis from the overhead view that precedes the incipient motion prior to a sequential swing motion of the kicking leg.

The results of the present study scientifically demonstrated that the support leg motion strongly contributed to achieving mature performance of instep kicking in football.

References

Ae, M., Tang, H., & Yokoi, T. (1992). Estimation of inertial properties on the body segments in Japanese athletes. In *Biomechanisms 11: Form, motion, & function in humans* (pp. 23–33). Tokyo: The University of Tokyo Press.

Andersen, T. B., & Dörge, H. C. (2011). The influence of speed of approach and accuracy constrain on the maximal speed of the ball in soccer kicking. *Scandinavian Journal of Medicine & Science in Sports*, 21, 79–84. doi: 10.1111/j.1600-0838.2009.01024.x

Apriantono, T., Nunome, H., Ikegami, Y., & Sano, S. (2006). The effect of fatigue on instep kicking kinetics and kinematics in association football. *Journal of Sports Sciences*, 24, 951–60. doi: 10.1080/02640410500386050

Asami, T., & Nolte, V. (1983). Analysis of powerful ball kicking. In H. Matsui, & K. Kobayashi (Eds.), *Biomechanics VIII-B* (pp. 695–700). Champaign, IL: Human Kinetics.

Barfield, W.R., Kirkendall, D.T., & Yu, B. (2002). Kinematic instep kicking differences between elite female and male soccer players. *Journal of Sports Science & Medicine*, 3, 72–79.

Davis, R., Ounpuu, S., Tyburski, D., & Gage, J. (1991). A gait analysis data collection and reduction technique. *Human Movement Science*, 10, 575–587.doi: 10.1016/0167-9457(91)90046-Z

Dörge, H. C., Andersen, T. B., Sørensen, H., & Simonsen, E. B. (2002). Biomechanical differences in soccer kicking with the preferred and the non-preferred leg. *Journal of Sports Sciences*, 20, 293–299. doi: 10.1080/026404102753576062

Feltner, M. E. & Nelson, S. T. (1996). Three-dimensional kinematics of the throwing arm during the penalty throw in water polo. *Journal of Applied Biomechanics*, 12, 347–372.

Inoue, K., Nunome, H., Sterzing, T., Shinkai, H., Ikegami, Y. (2014), Dynamics of the support leg in soccer instep kicking. *Journal of Sports Sciences*, 32, 1023–1032. doi: 10.1080/02640414.2014.886126

Isokawa, M., & Lees, A. (1988). A biomechanical analysis of the instep kick motion in soccer. In T. Reilly, A. Lees, K. Davies, & W. J. Murphy (Eds.), *Science & Football* (pp. 449–455) London: E & FN Spon.

Katis, A., & Kellis, E. (2010). Three-dimensional kinematics and ground reaction forces during the instep and outstep soccer kicks in pubertal players. *Journal of Sports Sciences*, 28, 1233–1241. doi: 10.1080/02640414.2010.504781

Kellis, E., & Katis, A. (2007). Biomechanical characteristics and determinants of instep soccer kick. *Journal of Sports Science & Medicine*, 6, 154–165.

Kellis, E., Katis, A., & Gissis, I. (2004). Knee biomechanics of the support keg in soccer kicks from three angle of approach. *Medicine & Science in Sports & Exercise*, 36, 1017–1028. doi: 10.1249/01.MSS.0000128147.01979.31

Lees, A., & Nolan, L. (1998). The biomechanics of soccer: a review. *Journal of Sports Sciences*, 16, 211–34. doi: 10.1080/026404198366740

Lees, A., Steward, I., Rahmana, N., & Barton, G. (2009). Lower limb function in the maximal instep kick in soccer. In: T. Reilly, & G. Atkinson (Eds), *Contemporary Sport, Leisure & Ergonomics*. (pp. 149–159) London: Routledge.

Levanon, J., & Dapena, J. (1998). Comparison of the kinematics of the full-instep kick and pass kicks in soccer. *Medicine & Science in Sports & Exercise*, 30, 917–927. doi: 10.1097/00005768-199806000-00022

Luhtanen P. (1988). Kinematics and kinetics of maximal instep kicking in junior soccer players. In T. Reilly, A. Lees, K. Davies, & W. J. Murphy (Eds.), *Science & Football* (pp. 441–448) London: E & FN Spon.

Nunome, H., Asai, T., Ikegami, Y., & Sakurai, S. (2002). Three-dimensional kinetic analysis of side-foot and instep soccer kicks. *Medicine & Science in Sports & Exercise*, 34, 2028–36. doi: 10.1249/01.MSS.0000039076.43492.EF

Nunome, H., & Ikegami, Y. (2005). The effect of hip linear motion on lower leg angular velocity during soccer instep kicking. In Q. Wang (Eds.), *Proceedings of the XXIIIrd Symposium of the International Society of Biomechanics in Sports* (pp. 770–772) Beijing: The People Sports Press.

Nunome, H., Ikegami, Y., Kozakai, R., Apriantono, T., & Sano, S. (2006a). Segmental dynamics of soccer instep kicking with the preferred and non-preferred leg. *Journal of Sports Sciences*, 24, 529–541. doi 10.1080/02640410500298024

Nunome, H., Lake, M., Georgakis, A., & Stergioulas, L. K. (2006b). Impact phase kinematics of instep kicking in soccer. *Journal of Sports Sciences*, 24, 11–22. doi: 10.1080/02640410400021450

Roberts, E. M., & Metcalf, A. (1968). Mechanical analysis of kicking. In J. Wartenweiller, E. Jokl, & M. Hebbelink (Eds.). *Biomechanics I* (pp. 315–319), Basel: Karger.

Rodano, R., & Tavana, R. (1988). Three-dimensional analysis of instep kick in professional soccer players. In: T. Reilly, J Clarys, & A. Stibbe (Eds.), *Science & Football II* (pp. 441–448), London: E & FN Spon.

Sprigings, E., Marshall, R., Elliot, B., & Jennings, L. (1994). A three-dimensional kinematic method for determining the effective of arm segment rotations in producing racquet-head speed. *Journal of Biomechanics*, 27, 245–54.

Wickstrom, R. L. (1975). Developmental kinesiology. *Exercise & Sports Science Reviews*, 3, 163–92.

Winter, D. A. (2009). *Biomechanics and motor control of human movement*, 4th ed., Hoboken: John Wiley & Sons, Inc.

3 The biomechanics of place kicking in Rugby Union

*Neil Bezodis, Alexandra Atack
and Stacy Winter*

1. Introduction

Place kicks provide a means through which to score points in Rugby Union. They are a prevalent feature of the modern game with 45% of the total points scored in 582 international matches over a recent 10-year period coming from place kicks (Quarrie & Hopkins, 2015). The outcome of a single place kick can also be the determining factor in the outcome of an entire match; in the sample of matches analysed by Quarrie and Hopkins (2015), 33 match results depended directly on the success of a place kick. Additionally, the result of 14% of all of the matches would have differed if the place kick success percentages of the two teams were reversed. In spite of this clear importance of place kicking in Rugby Union, biomechanical research into rugby place kicking remains relatively scarce in comparison to that undertaken in other football codes, in particular instep kicking in soccer. However, as the average success rate for the 6,428 place kicks analysed by Quarrie and Hopkins (2015) was 73%, there is clear scope for rugby place kicking technique research to support improvements in performance levels and ultimately to affect the outcome of matches.

Given the applied nature of the challenge of improving rugby place kicking performance, it is important that any research undertaken leads to outcomes of improved practice of coaches and kickers. The approach adopted in this chapter is therefore based around Bishop's (2008) Applied Research Model for the Sport Sciences (ARMSS), which aims to "guide the direction of research required to build our evidence base about how to improve performance" (Bishop, 2008, p. 253). The ARMSS provides a useful framework around which to seek enhanced understanding of, and subsequently the coaching and execution of, rugby place kick technique. Owing to the relative lack of existing research in rugby place kicking, it is imperative that the ARMSS is addressed from its first stage. This stage involves the identification of real-world issues faced by coaches and athletes and requires consideration of any relevant existing literature to determine the current state of the knowledge about rugby place kicking. Early discussions between researchers and coaches are important for the success of such applied research, particularly at this early stage (Bishop, 2008). This will yield a more holistic and applied understanding and it is therefore more likely that potential problems can be clearly defined, allowing relevant research questions that will contribute to performance to be formulated. The second stage of the ARMSS, which helps to move research questions forward from this initial understanding, is based on descriptive research. Access to descriptive data from high-level place kickers can be used to provide valuable insight to inform and direct future research which will extend the evidence base further and ultimately lead to performance improvements.

The aim of this chapter is therefore to integrate experiential knowledge from an expert coach with existing biomechanical evidence and empirical kinematic data from high-level performers to develop the current understanding of technical aspects of rugby place kicking. This integration of qualitative and quantitative data from high-level applied practice is intended to provide relevant preliminary evidence upon which future experiments, applied interventions, and coaching practice can be based. It is hoped that the findings presented in this chapter will stimulate future interest in specific aspects of the biomechanics of rugby place kicking, which will further develop the understanding and, most importantly, will improve performance.

2. Methods

2.1 Participants

Following institutional ethical approval, one male International Performance Coach (age: 47 years) and 14 male place kickers (mean ± SD age: 19 ± 1 years; mass: 87.2 ± 7.6 kg; height: 1.83 ± 0.05 m) provided written informed consent to participate. The coach held the highest-level Rugby Football Union coaching qualification with a specialism in place kicking. Collectively, he had 16 years of full-time coaching experience in professional rugby, including nine years with a kicking-specific role within an International Union. The 14 place kickers were all contracted to professional Rugby Union clubs and were current International representatives at the Under 20 age group level.

2.2 Qualitative methods

An Interpretative Phenomenological Analysis (Smith, 1996) was adopted as an in-depth qualitative approach to explore the coach's understanding of place kicking. Following a pilot interview with a Rugby Football Union Level 3 qualified coach to refine the interview guide, a semi-structured interview comprising open-ended questions was conducted with the International Performance Coach at a location of his choosing. Although there was a certain element of structure to the interview, the order of questioning was dependent on the coach's responses, with freedom allowed to express his thoughts and expand on his views (Smith, 2008). The interview was transcribed verbatim and read several times by two of the authors to gain a sense of the whole before data reduction. Preliminary comments and associations were noted to develop the authors' understanding and these notes were then used to identify the emergent themes which were aligned with quotations from the transcript. During the course of analysis, the two authors had extensive discussions on the transcriptions and emerging themes to help uncover any biases in the lead author's analytic approach (Winter & Collins, 2015). Finally, member checking was performed to ensure that the representation of the coach's experiences and the emergent themes were accurate and to offer the coach the opportunity to add any further information (Brocki & Wearden, 2006).

2.3 Quantitative methods

An 11-camera motion capture system (Vicon©, MX3) was used in an indoor laboratory to obtain three-dimensional kinematic data at 240 Hz from place kicks performed by the 14 participants. Data were collected over a two-and-a-half year period and all kickers were current members of the Under 20 International squad at the time they were analysed. All kickers wore their own moulded rubber boots and, following a self-directed warm-up, performed a series of place kicks using a size 5 Gilbert Virtuo match ball from a kicking tee of their preference. All kicks were of maximal effort into a net which was placed approximately 2 m in front of the tee. Kickers were asked to aim towards a vertical line suspended in the net which represented the centre of the goalposts. The global Y-axis was defined horizontally along a line from the centre of the ball to the centre of the vertical target line, the Z-axis was vertical and the X-axis was determined as the cross-product of the Y- and Z-axes. The tee and net were located in a position which ensured that the non-kicking foot landed near the centre of a force platform (Kistler, 9287BA, 960 Hz) which was covered and flush with the surrounding rubber floor. Eighty reflective markers were placed on the kicker at specific anatomical locations to define a 14 segment rigid body model using a CAST approach (Cappozzo et al., 1995) and 54 of these markers were retained to track the segmental kinematics during each kick (Figure 3.1). Six markers were also placed on the ball to enable determination of the ball centre of mass (CM). Marker trajectories were labelled using Nexus (v. 1.8, Vicon©) and data were then exported for processing and analysis in Visual3D™ (v. 5.02, C-Motion, Inc.).

Resultant ball velocity was determined from the first derivative of polynomial functions fitted through the first five frames of raw ball CM displacement data after ball contact. Raw data from all of the markers attached to the kickers were low-pass filtered at 18 Hz. The segmental kinematics were then reconstructed from these filtered marker trajectories using an evenly-weighted inverse kinematics approach (Lu & O'Connor, 1999) and segmental and joint angles were calculated using an XYZ Cardan rotation sequence. Segmental inertia data were obtained from de Leva (1996) and whole body CM location was determined using a summation of segmental moments approach. Support foot contact was identified where the vertical ground reaction force first exceeded 10 N. Specific discrete kinematic variables (Figure 3.2) that were deemed to relate to the coach's interview responses were then extracted and the mean values for each kicker across four kicks were calculated. For all variables of interest, the 14 data values (i.e. one from each kicker) were checked for normality ($p > 0.05$) using a Shapiro-Wilk test and the mean and standard deviation (SD) of the group-wide data were

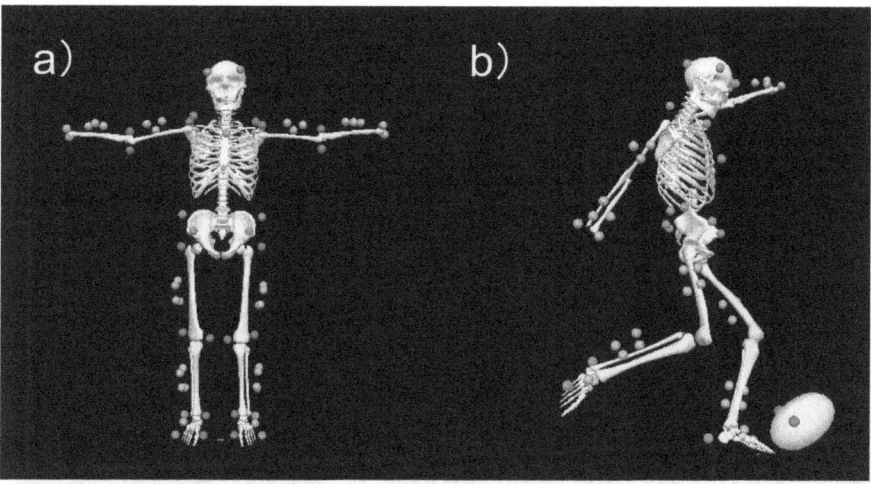

Figure 3.1 Illustration of the markers used to define the 14 rigid body segments in the static trial (panel a), and to track the segments during the dynamic kicking trials (panel b)

calculated. Maximum and minimum values from the group-wide data were also extracted to illustrate the extremes in the identified features of technique across this group of kickers of relatively homogeneous ability level.

3. Results and discussion

The interview lasted 57 minutes and the subsequent analysis of the coach's under-standing revealed two emergent themes of relevance to the aims of this chapter. The first theme corresponded to the general aims and philosophy of the coach who made it clear that, "the be all and end all in this profession is results". However, in addition to kicking success percentages – "anything above 80% is good" – it was also impor-tant for the coach to consider "the longevity of the player…he could kick eight out of eight one week and then not play for the next ten weeks; that defeats the object". Given the average success rate of 73% reported by Quarrie and Hopkins (2015), the achievement of 80% appears to be a suitable goal which sports biomechanists, working closely with coaches, clearly have the potential to support. Previous rugby place kicking research has focussed entirely on performance outcome, typically in terms of ball velocity, but it is clear from the coach's response that injury prevention is also a necessary consideration. The technical attributes subsequently identified by the coach should therefore be considered in the context of player welfare in addition to performance outcome, and future research should be clear about its focus while remaining mindful of this dual consideration.

The second emergent theme corresponded to the attributes associated with successful place kicking and was subdivided in to technical, mental, and physical factors. Given the aim of this chapter, the technical attributes were explored in detail

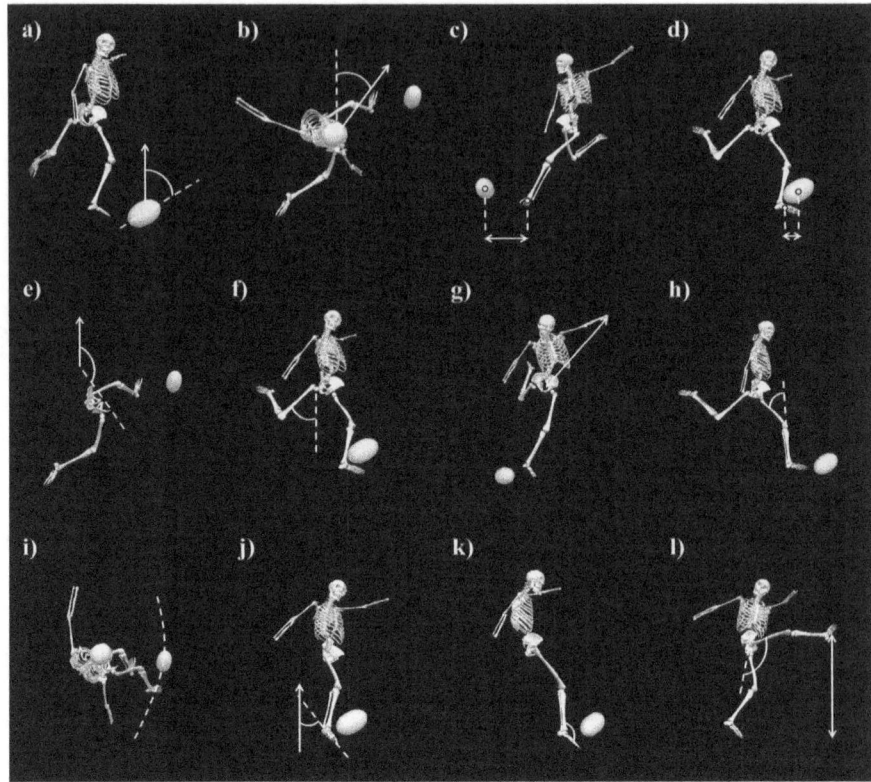

Figure 3.2 Visual identification of selected kinematic variables determined in the quantitative analysis. From left to right, top to bottom: a) ball angle; b) approach angle; c) medio-lateral support foot placement; d) antero-posterior support foot placement; e) pelvis angle at support foot contact; f) thigh angle at support foot contact; g) non-kicking-side wrist to pelvis distance at support foot contact; h) support leg knee flexion at support foot contact; i) kicking foot trajectory; j) kicking foot angle at ball contact; k) kicking leg ankle angle at ball contact; l) peak kicking leg hip flexion angle and peak kicking foot CM height during the follow through. Some segments are removed in Figures e, j and k to improve the visibility of the depicted kinematic variable

during the interview. A phase analysis (Lees, 2002) was applied to retrospectively divide these technical attributes in to different phases of the movement, namely the approach (all actions up to and including support foot contact), the kicking action (from support foot contact to ball contact), and the follow through (from ball contact onwards). The technical attributes which the coach identified within these phases were then used to guide the variables extracted (Table 3.1) from the quantitative analysis of the 14 high-level kickers. The ability level of these kickers was reinforced by the resultant ball velocities they achieved during the laboratory analysis (mean ± SD = 27.4 ± 1.9 m/s, range = 24.5 to 30.5 m/s). These values are higher than those previously reported from a group of professional kickers (n = 14, mean ± SD = 26.4 ± 3.0 m/s; Holmes *et al.*, 2006) and from groups of lower-level kickers (mean values ranging from 17.8 to 26.4 m/s; Baktash *et al.*, 2009; Bezodis *et al.*, 2007; Padulo *et al.*, 2013; Sinclair *et al.*, 2014; Zhang *et al.*, 2012).

Table 3.1 Selected discrete descriptive data from the group of high-level place kickers.

Variable	Mean	SD	Minimum	Maximum
Ball angle on tee (°)	34	17	2	56
Approach angle (°)	34	6	25	47
Medio-lateral distance between support foot CM and ball CM (m)	0.32	0.04	0.25	0.37
Antero-posterior distance between support foot CM and ball CM (m)	0.09	0.07	−0.02	0.19
Pelvis angle at support foot contact (°)	139	9	126	157
Kicking leg thigh angle at support foot contact (°)	−58*	n/a	−24	−71
Resultant distance between non-kicking-side wrist and pelvis CM at support foot contact (m)	0.97	0.09	0.84	1.08
Support leg knee flexion angle at support foot contact (°)	27	5	21	37
Total support leg knee flexion during support leg contact (°)	22	4	14	29
Medio-lateral range of motion of the kicking foot CM in the 0.3 m prior to initial ball contact (m)	0.11	0.02	0.07	0.14
Medio-lateral range of motion of the kicking foot CM in the 0.3 m after initial ball contact (m)	0.02	0.01	0.01	0.04
Kicking leg foot angle at initial ball contact (°)	46	8	32	61
Kicking leg ankle angle at ball contact (°)	25	6	15	37
Peak kicking leg hip flexion angle during follow through (°)	104	11	88	121
Peak height of kicking leg foot CM during follow through (% of standing height)	46	12	28	62

Ball angle: The angle of the long axis of the ball about the global medio-lateral (X) axis (see Figure 3.2a). 0° represents a vertically aligned long axis of the ball, positive angles represent the top of the ball anterior to the bottom of the ball.

Approach angle: the direction of whole body CM motion in the horizontal (X-Y) plane during the airborne phase immediately prior to support foot contact (see Figure 3.2b). 0° represents motion directly along the line towards the target (i.e. Y-axis), positive angles represent approaching from towards the non-kicking leg side.

Antero-posterior distance between support foot CM and ball CM: A negative value represents the foot CM ahead of the ball CM.

Pelvis angle: The angle of the pelvis about the global vertical (Z) axis (see Figure 3.2e). 90° represents the line between both ASIS being perpendicular to the line of the target (i.e. Y-axis), angles >90° represent the kicking side of the pelvis retracted further away from the target than the non-kicking side.

Thigh angle: The angle of the thigh segment about the global medio-lateral (X) axis (see Figure 3.2f). 0° represents a vertically oriented thigh segment, positive values represent the distal end of thigh rotated anteriorly relative to the proximal end. *The median value is presented because the Shapiro-Wilk test revealed data to not be normally distributed (note: all kickers exhibited values between −52° and −71°, aside from one kicker who had a value of −24°).

Foot angle: The angle of the foot segment about the global medio-lateral (X) axis (see Figure 3.2j). 0° represents a vertically oriented foot segment, positive values represent the distal end of the foot rotated anteriorly relative to the proximal end.

Kicking leg ankle angle at ball contact: Relative angle between the foot and shank segment (see Figure 3.2k), positive values represent plantar flexion relative to the angle in the static trial.

3.1 The approach

The coach identified that ball placement on the tee is largely individual, stating: "there's an advance for people leaning the ball slightly to open a sweet spot... some like [to kick] on the point...others [prefer to position the ball] upright". The quantitative data confirmed this as the orientation of the long axis of the ball on the tee about the global medio-lateral (X) axis ranged from 2° (near vertical) to 56° (top of the ball pointing towards the posts) between kickers (Figure 3.2a). Further analysis of the data revealed that five of the kickers positioned the ball between 53° and 56° ('kicking on the point'), while eight positioned the ball between 2° and 26° ('leaning the ball slightly') and only one of the 14 kickers used a ball angle (34°) between these two distinct set ups. Given that a rugby ball is not spherical, it is likely that these different orientations affect the foot–ball interaction at contact. Future high-speed analysis of this foot-ball interaction, similar to that previously performed for soccer kicking (e.g., Ishii et al., 2012), could yield valuable insight regarding the relative merits of different ball orientations and their potential interactions with the rugby place kicking techniques used.

All kickers adopted an angled approach towards the ball; the direction of CM displacement in the horizontal X-Y plane during the final airborne phase immediately prior to support foot contact ranged from 25° to 47° relative to the Y-axis (mean ± SD = 34 ± 6°; Figure 3.2b). While not explicitly discussed by the coach, this angled approach may serve to facilitate some of the specific technical features he subsequently identified as important, such as pelvic retraction on the kicking leg side at support foot contact (Figure 3.2e). Although experimental manipulations to approach angle were previously investigated in rugby place kicking by Padulo *et al.* (2013), only kicking knee angular kinematics were reported and no effects of different approach angles were found on these. The use of an angled approach by the kickers in the current investigation is consistent with that commonly observed in soccer instep kicking, where it has been proposed to influence pelvis kinematics at support foot contact (Lees *et al.*, 2010). The direction of the approach in rugby place kicking could therefore influence pelvis or upper body kinematics at support foot contact and this is an area worthy of future investigation, particularly given the coach's comments on the importance of body configuration at support foot contact which are addressed in the *kicking action* section of this discussion.

Upon landing from the final airborne phase, the support foot CM was placed just over 0.3 m from the ball CM in the global medio-lateral (X-axis, Figure 3.2c) direction. This was relatively consistent between kickers (mean ± SD = 0.32 ± 0.04 m) and the mean value matches the mean medio-lateral distance previously reported between the heel and ball from a separate group of 15 professional place kickers (Cockcroft & van den Heever, 2016). There was slightly more variation in the antero-posterior (Y-axis) foot placement (mean ± SD = 0.09 ± 0.07 m, Figure 3.2d); most kickers positioned their support foot CM behind the ball CM (up to a maximum of 0.19 m) although one kicker placed it directly in line and another placed the foot 0.02 m ahead of the ball CM. Support foot placement has previously been shown to be consistent both within and between kickers, particularly when compared with foot placement in the two prior approach steps

(Cockcroft & van den Heever, 2016). This relatively low inter- and intra-kicker variation in support foot placement suggests that previous experimental manipulations of 0.30 m in anterior, posterior, and medial directions in rugby place kicking (Baktash *et al.*, 2009) may have been excessive and therefore of limited ecological validity. However, given that the support foot is the most distal segment of the linked-segment kicker system and that it locates the kicker in the global space relative to the location of the ball, its placement is likely to play an important role in dictating the path of the kicking foot. As the coach subsequently attributed importance to the path of the kicking foot when discussing the kicking action, support foot placement should remain an important consideration when investigating rugby place kicking technique.

3.2 The kicking action (from support foot contact to ball contact)

The coach discussed the importance of a "triangle" between the support foot, the kicking foot and the non-kicking side shoulder at support foot contact. He suggested that, "the more stretched this is, the more power that should hopefully be generated". The "stretch" of the non-kicking side shoulder and kicking foot away from the support foot at contact could relate to a 'tension arc' which was previously identified in skilled soccer players performing instep kicks (Shan & Westerhoff, 2005). Shan and Westerhoff (2005) proposed that increased length of trunk and hip flexor musculature at support foot contact assisted the subsequent generation of larger concentric muscle forces, supporting the coach's suggestion that the "triangle [is] like an elastic band, stretch it". The current cohort of kickers landed with their pelvis retracted away from the posts on the kicking side at support foot contact (by $139 \pm 9°$ relative to the pelvis being parallel to the Y-axis, Figure 3.2e), a feature of technique which was likely enabled by the angled approach discussed previously. Hip extension was evident around support foot contact, serving to rotate the thigh segment away from the ball (mean $\pm$ SD absolute thigh angle about the global medio-lateral (X) axis $= -56 \pm 10°$, Figure 3.2f). This combination of pelvic retraction and hip extension therefore facilitated the "stretch" between the two feet in the "triangle" at support foot contact.

While the "stretch" of the non-kicking side shoulder from the feet could directly relate to the creation of a 'tension arc' (Shan & Westerhoff, 2005), it may also be an effect of movements of the non-kicking side arm. At support foot contact, the non-kicking side arm was abducted and horizontally extended such that the wrist was 0.97 ± 0.09 m ($53 \pm 5\%$ of total standing height) away from the pelvis (Figure 3.2g). This is a position from which the non-kicking side arm can generate considerable angular momentum during the kicking phase, a feature which has previously been identified as important for accurate rugby place kicking, particularly when also striving to achieve maximal distance (Bezodis *et al.*, 2007). The third point of the coach's "triangle", the non-kicking foot, was "stretched" away from other parts of the body through a relatively extended knee ($27 \pm 5°$ of flexion, Figure 3.2h) at support foot contact. The coach used the analogy of a rigid "pillar" for the non-kicking leg and suggested the importance of "locking out your [non-kicking] leg in order for your [kicking leg] to strike through". However, all kickers exhibited further support leg knee flexion (mean ± SD total range of flexion = 22 ± 4°) between support foot contact and ball contact. While completely eliminating knee flexion during the kicking action (i.e. maintaining a rigid "pillar") therefore appears to be an unlikely expectation based on the amount of flexion exhibited by these high-level kickers, it is plausible that limiting the amount of flexion could be an important feature of place kicking technique. Future research should therefore consider the importance of the support leg and assess the relative merits of technical or physical coaching methods designed to limit support leg knee flexion during place kicking.

From the "stretched" position at support foot contact, the coach made it clear that, "the hip leads the knee, the knee then snaps the shank that snaps the foot through the ball underneath it". Such proximal-to-distal sequencing is a widely accepted feature of striking skills such as kicking (Putnam, 1993). All of the analysed kickers exhibited clear proximal-to-distal sequencing in the linear velocities of the segment endpoints (i.e., hip, knee, ankle) and the segmental angular velocities (i.e., thigh, shank), confirming the existence of this sequencing in high-level rugby place kickers. In addition to this, the coach suggested that kickers could not be successful "if they haven't got any...coordination". Although it is beyond the scope of the preliminary descriptive nature of this chapter, intra-limb coordination remains relatively unexplored in kicking actions but may be a relevant direction for future research, particularly if the research questions are concerned with the learning and development of rugby place kicking technique (Chow *et al.*, 2007, 2008).

The coach proposed that the kicking foot should travel "in a straight line going through...towards the target...from six to twelve inches behind the ball [until]...six to twelve inches after impact". However, closer inspection of the raw trajectories of the kicking foot CM from the empirical data (Figure 3.2i) appeared to partly conflict this. Over the 0.3 m antero-posterior distance immediately behind the point of ball contact, the foot CM travelled in a curvilinear path, and the kickers exhibited a medio-lateral range of foot CM motion of 0.11 ± 0.02 m. This appears logical given the linked-segment nature of the system and that the path of the foot has been shown to follow an inclined plane during soccer kicking (Alcock *et al.*, 2012). However, in the 0.3 m immediately after the point of ball contact, the medio-lateral range of motion of the foot CM was considerably less (mean ± SD = 0.02 ± 0.01 m). This suggests that although the foot may not be travelling in a straight line towards the target immediately prior to ball contact, during the actual volume where ball contact is made, the foot path is near straight towards the target. Given the relatively homogeneous and high ability level of the studied kickers, this may be a feature of their skill level, particularly as this is likely to affect consistent, accurate kicking. Further investigation prior to, during and after ball contact is required to fully understand the importance of the path of the kicking foot. This could also assess the appropriateness and potential influence of the "foot towards target" principle, which was held by the coach and is a technical feature proposed to be important across several coaching texts (e.g., Biscombe & Drewett, 1998; Greenwood, 2003; Wilkinson, 2005). The coach also suggested that at ball contact, the foot should be "toe down...promote the hard part of the foot striking through the ball". This appears logical given the nature of impact mechanics and the desired outcome of a high ball velocity. The kicking foot segment of the kickers was angled at $46 \pm 8°$ relative to the global medio-lateral (X) axis at ball contact (Figure 3.2j), and this was partly achieved through $25 \pm 6°$ more plantar flexion than in a neutral standing position (Figure 3.2k). Given the previously identified variation in ball angle on the kicking tee, it would be prudent to investigate potential interactions between these foot kinematics and the initial ball angle, and this should be done using high-speed video given the relatively short duration of foot–ball contact.

3.3 The follow through

The coach suggested that the follow through is important because "there needs to be a…release mechanism…at the end…to dissipate the energy build up [due to] the forces they're putting on themselves". He also identified that the follow through is typically highly individual and that it may be "a hop or a skip, it may be a run, a step on your kicking foot afterwards. It may be whatever it is but there needs to be a release". Simple visual analysis of the 14 kickers confirmed that different strategies existed as nine kickers 'hopped' forwards to land on their non-kicking leg in the next ground contact after ball contact, while the remaining five kickers 'stepped' forward to take the next ground contact phase with their kicking leg. The kickers with the 'hopping' style exhibited peak kicking hip flexion ranging from 100 to 121° during the follow through, with the foot CM reaching a peak height of 42 to 62% of standing height (Figure 3.2l). In contrast, the kickers with the 'stepping' style follow through exhibited less peak kicking hip flexion (range = 88 to 93°) and a lower peak foot CM height (range = 28 to 38%). Given that the ball has left the foot, and thus the performance outcome of the movement has been determined by the time of the follow through, it has seldom been analysed in previous rugby kicking research, all of which has been performance outcome focussed. However, player longevity is also an important outcome consideration, particularly given the previously discussed comments in relation to the coach's aims and philosophy. The demands of the follow through on the musculo-skeletal system are therefore of interest for future research, particularly given the different follow through strategies evident within this relatively homogenous ability level group of kickers. Interestingly, it was recently demonstrated that experimental manipulations to the follow through could be useful when learning variations of simple motor skills such as grasping the handle of a robotic interface (Howard *et al.*, 2015). If these findings transfer to whole body skills such as kicking, follow through manipulations could be used in performance outcome focussed research in an attempt to experimentally affect movements during the kicking action phase.

4. Conclusion

The qualitative data obtained from the semi-structured interview with the coach yielded a rich, experiential understanding of rugby place kicking technique. These data were used to identify potentially important features of rugby place kicking technique during the approach, the kicking action and the follow through.

Three-dimensional kinematic data were then used to quantify these features of technique amongst a homogeneous group of high-level place kickers. This combination of qualitative and quantitative data has highlighted several features of technique, some of which may influence others, and many of which are worthy of future applied research. Two distinct styles of ball placement on the tee were identified and these may influence, or be influenced by, the foot kinematics at ball contact. The angle at which a kicker approaches the ball may also be of interest for understanding how it influences pelvis kinematics at support foot contact, and ultimately how the kicking leg segments are consequently affected. The location of support foot contact is also a potentially related consideration as it affects the location of these other segments, all of which combine to determine the path of the kicking foot before, during and after ball contact. Different styles in the follow through used after ball contact were also identified and the implications of these for both injury and performance should be investigated.

Clearly the biomechanical understanding of rugby place kicking remains in its relative infancy. This chapter aimed to provide an appropriate, evidence-based overview of some of the important features of technique and it is hoped that it will stimulate future experimental and theoretical biomechanical research to further the current understanding and ultimately to improve rugby place kicking performance.

Acknowledgements

The authors are grateful to Mr Jack Lineham for his technical assistance throughout the quantitative data collection.

References

Alcock, A. M., Gilleard, W., Hunter, A. B., Baker, J., & Brown, N. (2012). Curve and instep kinematics in elite female footballers. *Journal of Sports Sciences*, 30(4), 387–394.

Baktash, S., Hy, A., Muir, S., Walton, T., & Zhang, Y. (2009). The effects of different instep foot positions on ball velocity in place kicking. *International Journal of Sports Science and Engineering*, 3(2), 85–92.

Bezodis, N., Trewartha, G., Wilson, C., & Irwin, G. (2007). Contributions of the non-kicking-side arm to rugby place-kicking technique. *Sports Biomechanics*, 6(2), 171–186.

Biscombe, T., & Drewett, P. (1998). *Rugby: Steps to success*. Champaign, IL: Human Kinetics.

Bishop, D. (2008). An applied research model for the sport sciences. *Sports Medicine*, 38(3), 253–263.

Brocki, J. M., & Wearden, A. J. (2006). A critical evaluation of the use of interpretative phenomenological analysis (IPA) in health psychology. *Psychology and Health*, 21(1), 87–108.

Cappozzo, A., Catani, F., Della Croce, U., & Leardini, A. (1995). Position and orientation in space of bones: anatomical frame definition and determination. *Clinical Biomechanics*, 10(4), 171–178.

Chow, J. Y., Davids, K., Button, C., & Koh, M. (2007). Variation in coordination of a discrete multiarticular action as a function of skill level. *Journal of Motor Behavior*, 39(6), 463–479.

Chow, J. Y., Davids, K., Button, C., & Koh, M. (2008). Coordination changes in a discrete multi-articular action as a function of practice. *Acta Psychologica, 127*(1), 163–176.

Cockcroft, J., & van den Heever, D. (2016). A descriptive study of step alignment and foot positioning relative to the tee by professional rugby union goal-kickers. *Journal of Sports Sciences, 34*(4), 321–329.

de Leva, P. (1996). Adjustments to Zatsiorsky-Seluyanov's segment inertia parameters. *Journal of Biomechanics, 29*(9), 1223–1230.

Greenwood, J. (2003). *Total rugby: Fifteen man rugby for coach and player.* London, UK: Bloomsbury.

Holmes, C., Jones, R., Harland, A., & Petzing, J. (2006). Ball launch characteristics for elite rugby union players. In E. F. Moritz & S. Haake (Eds.), *The Engineering of Sport 6* (pp. 211–216), New York, Springer.

Howard, I. S., Wolpert, D. M., & Franklin, D. W. (2015). The value of the follow-through derives from motor learning depending on future actions. *Current Biology, 25*(3), 397–401.

Ishii, H., Yanagiya, T., Naito, H., Katamoto, S., & Maruyama, T. (2012). Theoretical study of factors affecting ball velocity in instep soccer kicking. *Journal of Applied Biomechanics, 28*(3), 258–270.

Lees, A. (2002). Technique analysis in sports: a critical review. *Journal of Sports Sciences, 20*(10), 813–828.

Lees, A., Asai, T., Andersen, T. B., Nunome, H., & Sterzing, T. (2010). The biomechanics of kicking in soccer: a review. *Journal of Sports Sciences, 28*(8), 805–817.

Lu, T.W., & O'Connor, J.J. (1999). Bone position estimation from skin marker co-ordinates using global optimisation with joint constraints. *Journal of Biomechanics, 32*(2), 129–134.

Padulo, J., Granatelli, G., Ruscello, B., & D'Ottavio, S. (2013). The place kick in rugby. *Journal of Sports Medicine and Physical Fitness, 53*(3), 224–231.

Putnam, C. A. (1993). Sequential motions of body segments in striking and throwing skills: descriptions and explanations. *Journal of Biomechanics, 26* (Suppl. 1), 125–135.

Quarrie, K. L., & Hopkins, W. G. (2015). Evaluation of goal kicking performance in international rugby union matches. *Journal of Science and Medicine in Sport, 18*(2), 195–198.

Shan, G., & Westerhoff, P. (2005). Full-body kinematic characteristics of the maximal instep soccer kick by male soccer players and parameters related to kick quality. *Sports Biomechanics, 4*(1), 59–72.

Sinclair, J., Taylor, P. J., Atkins, S., Bullen, J., Smith, A., & Hobbs, S. J. (2014). The influence of lower extremity kinematics on ball release velocity during in-step place kicking in rugby union. *International Journal of Performance Analysis in Sport, 14*(1), 64–72.

Smith, J. A. (1996). Beyond the divide between cognition and discourse: Using interpretive phenomenological analysis in health psychology. *Psychology and Health, 11*(2), 261–271.

Smith, J. A. (2008). *Qualitative psychology: A practical guide to research methods.* London: Sage Publications.

Wilkinson, J. (2005). *How to play rugby my way.* London: Headline Book Publishing.

Winter, S., & Collins, D. (2015). Why do we do, what we do? *Journal of Applied Sport Psychology, 27*, 35–51.

Zhang, Y., Liu, G., & Xie, S. (2012). Movement sequences during instep rugby kick: a 3D biomechanical analysis. *International Journal of Sports Science and Engineering, 6*(2), 98–95.

4 Impact in ovoid ball kicking

Kevin Ball

1. Introduction

Impact between the foot and the ball is the key to kicking performance. Everything that happens before impact is an attempt to optimize impact conditions such as the speed, movement direction and angle of the foot. Once the ball has left the boot, there is nothing the kicker can do to influence the outcome. This makes impact and the ability to manipulate it, the most important aspect of kicking.

Of the six football codes worldwide, four use ovoid shaped balls: Australian football, rugby league, rugby union and American football. In all four sports, kicking forms an important part of the game. In Australian football, kicking is the major form of passing and the only way a goal can be scored. In the rugby codes and American football, the place kick is used to score points from penalties or converting tries or touchdowns. Punt kicking is also performed in these codes to gain field position and is used as an attacking strategy in the rugby codes. Kicking the ovoid ball presents a different task to that of the spherical ball as impact characteristics will differ when contacting different parts of the ball (Nunome *et al.*, 2014). For example, the surface shape of these balls differ from the 'belly' of the ball that represents a wide and relatively flat surface, compared to the pointed and smaller area of each end of the ball.

Exploring the features of impact in kicking is an important research direction in studying kicking. It is also an essential area of the kick for coaches to understand and use in coaching practices. The emergence of ultra high-speed cameras has made this examination more realistic and achievable from both a research and a coaching viewpoint. This chapter will overview and evaluate findings of impact in the sports using an ovoid shaped ball.

2. Impact and performance

Impact characteristics for kicking an ovoid ball have been reported for Australian Football and Rugby League (Table 4.1). The methods for assessing impact have involved using high-speed video of between 1000 and 6000 frames per second with the video view focused on impact. Markers on the ball and foot have been tracked through impact to examine the change in velocity and to calculate instantaneous force. Initial and final conditions have also been calculated using similar methods to identify the foot and ball velocity immediately before and immediately after contact. Combining this data with contact time and distance (the time the foot and ball are in contact with each other and the distance the ball moves during this time) parameters such as average force (F = ma where m = mass of the ball and a = acceleration calculated as the change in velocity of the ball divided by contact time), and work (force times contact distance) are calculated.

Impact in kicking an ovoid ball is a forceful activity performed over a short duration. Average forces of over 1000 N have been reported with this force being applied for between 7–13 m/s depending on the kick type. The ball moves between 12–23 cm while in contact with the boot. Foot to ball speed ratios have been between 1.16 to 1.31 for different kick types or skill levels. Location of impact on the ball varies from near the point for place kicks (also a location some rugby code players choose for punt kicks), to one third up the ball for drop punt kicks (kicks that spin backwards) to the middle of the ball for torpedo kicks or flat punts (Figure 4.1).

Differences have been found for impact characteristics for kicks over different distances. Ball (2009a) compared 30 m and 50 m drop punt kicks with 1000 Hz video, finding greater foot and ball speeds as well as more work being done on the ball in the longer kick (Table 4.1). This is not surprising given the need to generate greater ball speed to achieve the increased distance of the kick. Peacock *et al.* (2017)

Table 4.1 Summary of impact research involving ovoid balls.

	Sport	Kick Task	Age/Skill Level	Foot Speed (m/s)	Ball Speed (m/s)	Foot to ball speed ratio	Contact time (m/s)	Contact Distance (m)	Average Force (N)	Work (J)
Ball	AF	50 m	Senior				10.0	0.24		271
(2009a)		30 m	Elite				9.8	0.19		198
Smith *et al.*	AF	Max	Senior	26.5	32.6	1.23	11.5	0.22		225
(2009)		Max NP	Elite	22.6	27.0	1.20	12.0	0.19		156
Ball (2010)	AF	Max	16–17 yrs. Elite	21.3	24.7	1.16	11.1	0.20	679	136
Peacock *et al.*	AF	Max	Senior	22.1	28.1	1.28	12.1	0.23	834	190
(2017)		Accuracy	Elite	17.7	22.1	1.25	13.2	0.20	558	119
Ball *et al.*	RL	Max	Senior	21.3	27.8	1.31	10.9	0.18	1045	
(2013)			Elite							
Ball (2010b)	RL	45 m	Senior	20.0	25.8	1.30	7.2	0.20		290
		Bomb	Elite	21.0	26.9	1.28	6.8	0.20		342
		Grubber		11.0	13.7	1.27	8.8	0.12		68
		Drop Kick		21.8	26.5	1.27	7.1	0.23		316
		Place Kick		21.2	25.2	1.20	7.4	0.22		306

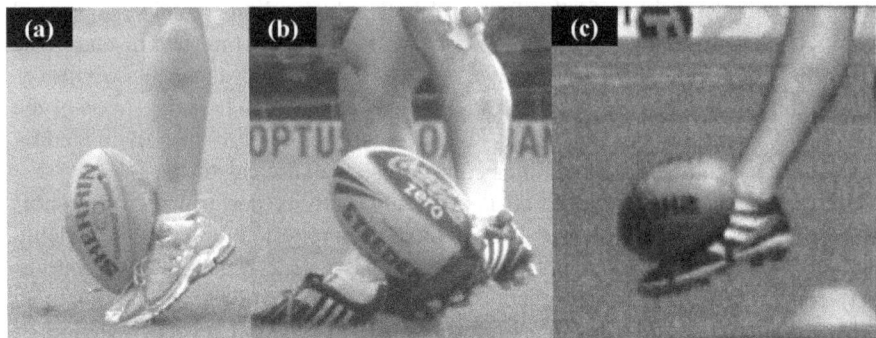

Figure 4.1 Different impact locations for different kicks: (a) Drop punt, (b) place kick and
       (c) torpedo

also found greater foot and ball speeds in comparing maximal kicks (approximately 60 m) with accuracy kicks over 20 m. However, it was interesting to note that even though kick distance was reduced by two thirds from 60 m to 20 m, players did not reduce their foot speed by a similar ratio. Peacock *et al.* (2017) suggested that the reason for this might be to produce a flatter trajectory that effectively increases the target area (Figure 4.2). This highlights that kicks over the same shorter distance can be performed in multiple ways from a slow lobbed kick to a fast 'spearing' kick, changing ball speed and flight time.

Differences in impact characteristics have been found between the preferred and non-preferred leg kicks (Smith *et al.*, 2009). This is potentially important in training the non-preferred leg, which does not perform as well as the preferred leg

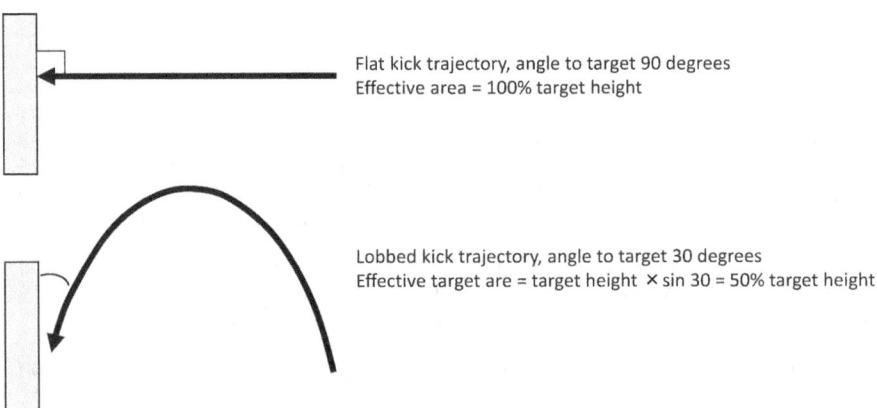

Figure 4.2 Flatter trajectory effectively retaining the full area of the target compared to a
       chipped kick that has a lower effective target area

in terms of distance, kicked and accuracy achieved. Senior elite AF players exhibited slower foot and ball speeds on their non-preferred leg kicks but interestingly, although the foot to ball speed ratio was also slightly lower (1.20 compared to 1.23), this was not significant. This indicated foot speed was the determining factor in the lower distances achieved, and not the nature of impact. With this lower foot speed, the ball was likely to be less deformed and this in turn led to a shorter contact distance, reducing the work done on the ball. So the focus of training to increase foot speed in non-preferred leg kicks would seem to be a priority.

Differences have been found between senior and junior performers (Ball *et al.*, 2010). Juniors produced lower foot and ball speeds compared to senior AF players. One reason for this was the significantly lower force applied to the ball by the juniors (junior players = 679 N, senior players 1023 N, calculated from data from Smith *et al.*, 2009). This 34% lower force as well as a lower contact distance meant that substantially less work was done on the ball. Foot to ball speed ratio (1.16 compared to 1.23) was not statistically significant and while the 6% difference might be of practical significance, it is not nearly as influential as the force applied to the ball. Juniors were also unable to hold their ankle as rigid as seniors, exhibiting 6 degrees of plantar flexion during ball contact compared to only 4 degrees for seniors (Table 4.2). This means the transfer of energy from the foot to the ball is less efficient and so a reduced ball speed will result in a loss of energy transfer to the ball and subsequently lower ball speed. The authors suggested that training to improve foot speed, force application on the ball and increasing ankle rigidity would assist junior players in gaining more distance.

While the decreased plantar flexion in seniors compared to juniors was indicative of better performance for distance, the ankle operates differently for shorter distance accuracy-based kicks. Elite senior AF players actually increased their plantar flexion in the shorter kicks from 2.2 to 7.2 degrees for the shorter kicks (Peacock *et al.*, 2017). Given forces were substantially lower for the shorter kicks, this can only be explained by an active (and not passive) 'softening' of the ankle. This might have been assisting the players to attain a more vertical foot angle and lower trajectory (foot angle was significantly more vertical and trajectory lower for accuracy kicks compared to distance kicks). It might also be linked to reducing high-pressure force gradients during contact, a factor suggested to be

Table 4.2 Ankle plantar flexion during impact.

	Sport	Task	Age/Skill Level	Average Force (N)	Change in Ankle Angle (°)
Smith *et al.* (2009)	AF	Max	Senior Elite	1023*	4*
Ball (2010)	AF	Max	16–17 yr Elite	679	6
Peacock *et al.* (2017)	AF	Max	Senior Elite	834	2.2
		Accuracy		558	7.2

* Calculated in the Ball (2010a) study for comparison with junior players

potentially important in soccer kicking for accuracy (Hennig & Sterzing, 2010). The underlying mechanism for this finding requires more research to substantiate but it is clear that different movement strategies are being used for distance and accuracy kicks.

3. Conditioning for impact

The large forces that exist during impact highlight the importance of conditioning for kicking. With contact time being only one hundredth of a second, and range of motion at the knee being approximately 15 degrees, manipulation of kicking itself offers the best approach to improving kick-specific strength at impact. Further, knee angular velocities of approximately 1500 degrees per second have been reported for punt kicking (Ball, 2008) but even the fastest isokinetic movements rarely exceed 300 degrees per second (20% of the kick values). This issue was highlighted by both Nunome *et al.* (2006) and was suggested as a reason for why strength and kicking research has been equivocal in its findings.

Progressive overload is an important conditioning principle and this can be achieved for kicking impact in a number of ways. Increasing kick length is an obvious one, although as has been established, players can still kick a ball with more force than is necessary simply to reach the target, so control of loads using this method can be an issue for shorter distances. Altering the mass of the ball being kicked is a simple method to change impact forces. This might be done by using balls of different size or, as was performed by Ball (2009b), soaking a leather ball in water (the ball absorbed some of the water and became heavier). Finally, different inflation pressures in the ball might vary loading patterns and represents an easy manipulation to use in training.

Use of overweight balls has been found to improve kicking distance. Ball (2009b) performed a structured kick distance programme using balls that were heavier than the regular game ball, finding improvements of 5 m in distance. Elite AFL players performed maximal distance kicks before being allocated to one of three groups: a distance kick group using weighted balls, a distance kicking group using regular mass balls and a control group. Ball weight was increased by 10–15% by soaking it in water immediately prior to the training session with a 2:1 kick ratio of weighted balls to regular mass balls used over eight sessions in four weeks. Both maximal distance groups improved but the weighted ball group improved slightly more than the regular ball group. Ball (2009b) suggested that had the intervention period been longer, the weighted ball group might have showed greater improvements. It might also be more influential to less trained players as the group tested was elite and were already highly trained. Impact was not specifically examined, but the increased mass would have increased the impact force required for maximal kicks and may have assisted in developing a stiffer ankle, although these factors need to be examined further to substantiate them.

Manipulation of impact point, ball size and inflation pressure can also change impact force in kicking. These manipulations might be best targeted at developing strength in junior players or providing a more structured rehabilitation process for

Table 4.3 Impact characteristics for different ball sizes, inflation levels and impact points.

	Foot:Ball Speed Ratio	Contact Time (m/s)	Contact Distance (m)	Average Force (N)	Max Force (N)	Work (J)
Size 5 Drop Punt	1.25	10.9	0.15	887	2364	132
Size 5 Drop Punt (5 psi)	1.21	13.4	0.18	698	1904	124
Size 4 Drop Punt	1.29	10.6	0.14	768	1962	109
Size 4 Drop Punt (5 psi)	1.27	13.0	0.18	592	1582	105
Size 3 Drop Punt	1.28	10.5	0.14	685	1749	97
Size 3 Drop Punt (5 psi)	1.25	12.7	0.17	507	1294	85
Size 2.5 Drop Punt	1.27	10.0	0.14	719	1805	100
Size 2.5 Drop Punt (5 psi)	1.24	12.7	0.17	549	1451	91
Size 5 Place Kick	1.32	11.8	0.17	869	2340	151
Size 4 Place Kick	1.37	11.8	0.17	707	1887	123
Size 2.5 Place Kick	1.38	11.5	0.18	621	1597	110

seniors coming back from injury. In recent experimental work in the ISEAL laboratory, a mechanical leg has been developed to quantify the effect of ball size and inflation pressure on impact in a controlled and methodical way. To explore the effects of ball size, pressure and impact location, four Gilbert rugby balls (size 2.5, 3, 4 and 5) were kicked at two inflation pressures (5 psi and 10 psi) and at two impact locations (mid ball replicating a drop punt as Figure 4.1(a) and on the point replicating a place kick as Figure 4.1(b)). For each ball and pressure condition, balls were placed on a Moose tee that was positioned on a platform attached to a kick leg system. Foot speed was set at 17 m/s, which approximately equates to a 30 m kick.

Ball size, inflation and impact location all affected impact characteristics (Table 4.3). Overall, a reduction in ball size reduced average and maximum force, contact time and work, while contact distance did not change substantially. A reduction in inflation also saw a decrease in maximum and average force as well as work. However, these changes were larger than those produced by changing ball size. Further, contact time and distance both increased. Comparing impact points, place kicks exhibited greater foot to ball speed ratios, contact times, distances and work indicating that impact location is also influential to impact characteristics in kicking.

The exception to the overall trend was the size 2.5 ball, indicating that ball shape also influences impact characteristics. This ball produced larger average and maximum forces and work compared with the size 3 ball in spite of being smaller in mass. This indicated that ball shape as well as mass is influential to impact as the size 2.5 ball was narrower and more pointed compared with the size 3 ball.

Including all ball conditions together (Table 4.4) provides a rank order from the highest to lowest force. The full-sized balls provided the largest average impact force values with the drop punt slightly higher than the place kick. At the lower

Table 4.4 Average force rankings and percentage of Size 5 drop punt kicks for all balls tested.

	Average Force (N)	% of Full size Punt kick
Size 5 Drop Punt	887	
Size 5 Place Kick	869	98%
Size 4 Drop Punt	768	87%
Size 2.5 Drop Punt	719	81%
Size 4 Place Kick	707	80%
Size 5 Drop Punt (5 psi)	698	79%
Size 3 Drop Punt	685	77%
Size 3 Place Kick	664	75%
Size 5 Place Kick (5 psi)	663	75%
Size 2.5 Place Kick	621	70%
Size 4 Drop Punt (5 psi)	592	67%
Size 3 Drop Punt (5 psi)	507	57%
Size 3 Place Kick (5 psi)	500	56%

Note: Size 2.5 place kicks at 5 psi were not performed

end, the size 3 ball deflated to 5 psi required the lowest impact force at 56–57% of the full-sized ball. This highlights the capacity to provide different loadings by manipulating ball size, inflation and impact location and to plan for and provide a progressive overload for kicking conditioning.

There were a number of other considerations in this evaluation. First, newer balls tended to produce higher forces. Second, increasing and reducing kick length can expand the continuum of impact forces presented here so can be used in conjunction (with the caveat that shorter kicks can still be struck with high force). Third, different inflation pressures can add increments to the progression. While two pressures have been presented here, pumping balls to different pressure can alter the relationship and provide more progression steps. Fourth, while force is important for load progression, the skill aspect of kicking needs to be considered as well. While under-inflated balls drop loads, too much kick volume with higher contact times and distance may affect game kick performance. A solution here is using smaller sized balls during later stages of the rehabilitation process to provide contact times and distances typical of game day kicks while using larger balls at lower pressures to provide greater impact force.

4. Conclusion

Impact in kicking an ovoid ball is highly forceful and occurs over a very brief time. Average forces have been reported to be over 1000 N with contact times between 12 and 23 m/s and contact distances of 0.1–0.2 m. Differences in impact characteristics have been found between kick distances, preferred and non-preferred kick legs, seniors and juniors, and between kick types. Impact characteristics are influenced by the size of the ball being kicked, the inflation pressure and impact point and these can be implemented to provide progressive overload when conditioning for kicking. Knowledge of impact characteristics is an important area to develop further for players, coaches and biomechanists.

References

Ball, K. (2008). Biomechanical considerations of distance kicking in Australian rules football. *Sports Biomechanics, 7*, 10–23. doi: 10.1080/14763140701683015

Ball, K. (2009a). Foot interaction during kicking in Australian rules football. In: T. Reilly & F. Korkusuz (Eds.), *Science and Football VI*, Abingdon: Routledge, pp. 36–40.

Ball, K. (2009b). Use of weighted balls to improve kicking distance. In: T. Reilly & F. Korkusuz (Eds.), *Science and Football VI*, Abingdon: Routledge, pp. 285–289.

Ball, K. (2010). Kick impact characteristics for different rugby league kicks. In: R. Jensen, W. Ebben, E. Petushek, C. Richter & K. Roemer (Eds.), *Proceedings of the 28th International Conference on Biomechanics in Sports,* Michigan State University, pp. 458–461. Retrieved from http://www.isbs.org/

Ball, K., Ingleton, C., Peacock, J., & Nunome, H. (2013). Ball impact dynamics in the punt kick. In: T-Y. Shiang, W-H. Ho, C. Huang & C-L. Tsai (Eds.), *Proceedings of the 31st International Conference on Biomechanics in Sports*, Taipei, Taiwan. Retrieved from http://www.isbs.org/

Ball, K., Smith, J., & MacMahon, C. (2010). Kick impact characteristics of junior kickers. In: T-Y. Shiang, W-H. Ho, C. Huang & C-L. Tsai (Eds.), *Proceedings of the 28th International Conference on Biomechanics in Sports,* Marquette: Michigan State University. Retrieved from http://www.isbs.org/

Hennig, E. M., & Sterzing, T. (2010). The influence of soccer shoe design on playing performance: a series of biomechanical studies. *Footwear Science, 2*(1), 3–11.

Nunome, H., Ball, K., & Shinkai, H. (2014) Myth and fact of ball impact dynamics in football codes. *Footwear Science, 6*, 105–118, doi: 10.1080/19424280.2014.886303

Nunome, H., Ikegami, Y., Kozakai, R., Apriantono, T., & Sano, S. (2006). Segmental dynamics of soccer instep kicking with the preferred and non-preferred leg. *Journal of Sports Sciences, 24*, 529–541.

Peacock, J., Ball, K., & Taylor, S. (2017 online). The impact phase of drop punt kicking for maximal distance and accuracy. *Journal of Sports Sciences* http://dx.doi.org/10.108 0/02640414.2016.1266015

Smith, J., Ball, K., & MacMahon, C. (2009). Foot-to-ball interaction in preferred and non-preferred leg Australian rules kicking. In: A. J. Harrison, R. Anderson & I. Kenny (Eds.), *Proceedings of the 30th International Conference on Biomechanics in Sports,* Limerick, Ireland. Retrieved from http://www.isbs.org/

5   Biomechanical characteristics of the long throw-in

Hironari Shinkai

1. Introduction

The soccer throw-in is a method of restarting play from the touchline using a player's hands. There are some detailed rules for delivering the ball in this restart. A player must face the field of play, have part of each foot on the touchline or on the ground outside the touchline, throw the ball with both hands from behind and over the head (The International Football Association Board (IFAB), 2016). To perform this technique with the ball in both hands makes this a distinct throwing technique when compared with other ball games. These rules restrict the free motion of the thrower thereby limiting the resultant throwing distance in most cases.

Rory Delap, an Irish professional footballer who played in the English Premier League, is one of the world's most famous long throwers in soccer, and was often referred as the "Human Sling." He could throw the ball with a flatter trajectory, and his throwing distance was up to 40 m, which could reach the far goal post from the touchline. His throw-in is, therefore, comparable to a free kick or corner kick as a strong weapon to set up a goal. Besides him, a female Irish professional footballer, Megan Campbell, is also very famous as a long thrower. In the 2015 FIFA Women's World Cup qualification, she assisted in two goals against Germany with a long throw-in during the game. It is evident from her video footage that she is capable of throwing a comparable distance to that of Rory Delap. Moreover, in the 2015/16 season, an amateur high-school footballer, who can throw the ball over 35 m, became very well known in Japanese soccer. It may be assumed that great upper body muscular strength is an essential factor for the long throw-in. However, as female and teen footballers can perform the long throw-in, muscular strength may be important but not a necessity; technical factors may be more important for a long distance throw-in.

Acar *et al.* (2009) reported the effectiveness of throw-ins during the 2006 FIFA World Cup in Germany (64 matches); the number of goals from a throw-in was 6, which corresponded to 4% of all 147 goals. Moreover, as mentioned above, the long throw-in is an effective play for scoring a goal, especially for teams with taller players. The long throw-in is also an important technique not only for the attacking final third, but also for other tactical situations. When a thrower is under intense pressure from the opponent players in their defensive third, the thrower often needs to throw the ball a long distance to clear that difficult situation. Hence throwing the ball far is an important ability for soccer players, especially the full-backs or wingers.

In contrast with the above, training for long throw-ins seems to have little focus in spite of its usefulness, because coaching cues derived from biomechanical research is lacking. Therefore, this chapter introduces the results of previous studies and one latest study on the three-dimensional characteristic motion of the upper body in a long thrower.

2. Literature review of technical factors

The soccer throw-in can be divided into three styles: a standing throw with both feet parallel, with feet staggered, and a throw with a run-up (with the exception of an unusual handspring throw-in). To produce the large vertical and anterior velocity of the ball, the throwers firstly utilize a countermovement, and then move their upper body forward. These actions can be characterized by four key events of: (a) retraction of the ball behind the head by trunk extension, shoulder flexion, and elbow flexion, (b) maximum ball retraction, (c) trunk and shoulder forward extension to bring the ball forward, (d) elbow extension to ball release (Lees, 2013). This proximal-to-distal sequence of segmental rotations has also been observed during other ball throwing actions.

The scientific papers that have analysed the throw-in motion are listed in Table 5.1. Almost all previous research has analysed the motion characteristics of the long throw-in action, whereas there is only one study which has reported motion data for an accuracy-based throw-in (Barraza & Yeow, 2015). Moreover, studies reporting ball trajectory including the initial release parameters are also listed in Table 5.2. The findings for longer throw-ins from these studies can be summarized as follows.

- The most important factor is ball velocity at the release.
- Ball landing distance of the run-up throw-in is longer than that of standing throw-in due to an increased approach velocity.
- The key motions for producing a large ball velocity are trunk flexion, and shoulder and elbow extension before ball release.
- The optimum release angle is approximately 30°.
- Ball release height has little influence on ball distance.
- Large backspin increases the ball distance.

The most essential factor to produce faster ball velocity is an increase in hand velocity at ball release. As described above, several researchers have reported shoulder flexion/extension and elbow flexion/extension as the motions which increase hand velocity. Previous researchers have only analysed these joint motions within the sagittal plane, even though the shoulder and elbow joint motions have multiple degrees of freedom. Furthermore, the participants in most of the studies are not habitual long throwers (i.e. < 30 m). Thus, a three-dimensional biomechanical understanding of the motions required for the long throw-in is necessary to further the current knowledge.

Table 5.1 Review of the literature reporting on human motion of long throw-ins and accurate throw-ins.

Authors (publication year)	Subject	Methods	Findings
Chang, G. (1979)	1 experienced footballer	• A camera (64 Hz) • Four types of throw: standing with parallel stance, staggered stance, run-up throw, handspring throw • Whole body kinematics and kinetics, and ball release parameters	• Largest ball velocity (18.3 m/s), distance (35.4 m), and release angle (45°) in handspring throw. • Larger ball velocity (15.2 m/s), release angle (45°), and ball distance (25.3 m) in run-up throw compared to those of the other two standing throw. • Most contribution to ball distance of the approaching velocity of the body and ball velocity.
Levendusky, T. A. et al. (1985)	12 male collegiate footballers	• A camera (100 Hz) and a force plate • Staggered stance throw with short approach • Upper body kinemtics in the sagittal plane and GRF	• Ball velocity (18.3 ± 1.2 m/s), release angle (29.2 ± 4.9°), ball distance (23.1 ± 3.4m). • Sequential movement patterns of the upper body. • Role of lead foot in stopping the moving body and translation the momentum to the upper body.
Messier, S. P. and Brody, M.A (1986)	13 male collegiate footballers (conventional throw) 2 collegiate and 2 youth footballers (handspring throw)	• A high-speed camera at 100 Hz (conventional throw) and 200 Hz (handspring throw) • Whole body kinematics and kinetics in the sagittal plane	• Ball velocity (18.1 m/s), release angle (28.1°), ball distance (29.3 m) in conventional throw. • Ball velocity (23.4 m/s), release angle (23.5°), calculated ball distance (44.2 m) in handspring throw of collegiate footballers. • Contribution to performance improvement of Rapid trunk flexion and shoulder and elbow extension just prior to release in conventional throw-in.
Kollath, E. and Schwirtz, A. (1988)	13 male collegiate footballers	• A camera (60 Hz) and a force plate • Standing and run-up throw • Whole body kinematics and kinetics in the sagittal plane	• Ball velocity (14.2 ± 1.3 m/s), release angle (33.0 ± 6.9°), ball distance (20.9 ± 2.9 m) in standing throw. • Ball velocity (15.3 ± 1.6 m/s), release angle (32.3 ± 5.7°), ball distance (24.1 ± 4.7 m) in run-up throw. • Correlation between ball distance and ball velocity, hand velocity, and approach velocity in run-up throw. • No significant correlations between any body angles and ball distance.

Study	Methods	Results
Lees, A. et al. (2005) 10 adult male footballers	• Motion capture system (240 Hz) • Staggered and run-up throw • Upper limb kinematics and kinetics in the sagittal plane	• Ball velocity (13.7 ± 0.8 m/s), release angle (14.3 ± 3.1°), ball distance (13.6 ± 1.8 m) in staggered throw. • Ball velocity (14.9 ± 0.8 m/s), release angle (12.6 ± 3.1°), ball distance (14.4 ± 1.5 m) in run-up throw. • Large accelerating torque by shoulder than that of elbow before ball release. • A significant relationship between shoulder retraction torque during pulled back phase and following propulsion torque for the run-up throw (r = −0.539).
Cerrah, A. O. et al. (2012) 6 male amateur footballers	• A camera (25 Hz) and a force plate • Standing throw with one step and run-up throw with three steps • 3D GRF of the forefoot	• Longer ball distance in run-up throw (21.4 ± 1.2 m) than that of standing throw-in (19.5 ± 1.9 m). • Larger vertical force of standing throw from back swing to ball release than that of run-up throw. • Shorter time of run-up throw in all phases leading to improvement of ball distance than that of standing throw. • Not determining factor of mediolateral force for throw-in performance, and important factor of anteroposterior force before release.
Barraza, L. C. H. and Yeow C. H. (2015) 12 male subjects	• Motion capture system (100 Hz) and two force plates • Standing and run-up throw with two steps at objective board 5.5 m ahead • Whole body kinematics and GRF • Differences in dominant and nondominant body side motions between skilled and less-skilled subjects	• Lesser throwing deviation of skilled subjects corresponding to almost half of that of less-skilled subjects. • Higher peak vertical force at the non-dominant support leg for skilled subjects in both throw-in styles. • Differences of hip and knee angular motion between throw-in styles or subject groups. • Importance of lower body motions and GRF for accurate throw-in.

Table 5.2 Review of the literature reporting on ball trajectory and ball release parameters of long throw-ins.

Authors (publication year)	Subject	Methods	Findings
Bray, K. and Kerwin, D. G. (2004)	2 male footballers (one is long throw specialist) A ball launching machine	• Conventional maximum throw (human subjects) and comparable launches (machine) • Mathematical model including the aerodynamic parameters calculated from the experimental data (Kerwin D. G. and Bray K., 2004)	• Confirmation of presence of ball backspin in actual throw. • More accurate modeling of ball trajectory by using drag and lift coefficient. • A decrease of optimum release angle (8°) with consideration the backspin (lift force). • Importance of ball velocity, release angle, and ball spin as key parameters for achieving maximum ball distance.
Kerwin, D. G. and Bray, K. (2004)	2 male footballers (one is long throw specialist) A ball launching machine	• Four cameras for trajectory and three cameras for ball spin (50 Hz) • Conventional maximum throw (human subjects) and comparable launches (machine) • Ball trajectory, release parameters, and landing position	• Plotting the whole of ball trajectory. • Release parameters (ball velocity, ball distance, maximum height, height at the release, release angle, spin rate) including a long thrower over 30 m. • Developing a video-based method to quantify the ball trajectory.
Linthorne N. P. and Everett D. J. (2006)	1 collegiate footballers	• A camera (50 Hz) • Maximum throw after short run-up 0-4 strides with various release angles • Ball flight modeling using release parameters of the ball	• Ball velocity (13.4±0.3 m/s), release angle (32.1±1.6 °), ball spin rate, and ball distance (17.0 ± 0.8 m) in preferred release angle condition. • Calculated optimum release angle is approximately 30° considering musculoskeletal structure of the player's body. • Slight difference of release angle (around 30°) has small effect on maximum achievable distance. • Release height has small effect on the optimum release angle. • Effectiveness of high backspin to increase the ball distance.

3. Three-dimensional characteristic motions

To address the deficiency of in previous research, a three-dimensional analysis of the upper body motion of a group of throwers including one outstanding long thrower was conducted (Shinkai, 2014). Fourteen collegiate footballers (age 20.7 ± 1.4 yrs; height 176.1 ± 4.5 cm; body mass 70.7 ± 10.3 kg; career 11.9 ± 2.5 yrs) participated in this study. The experiment was conducted indoors on artificial turf to avoid the effects of wind. Participants were instructed to perform several throws of the ball as far as possible with a free run-up in accordance with the rules (IFAB, 2016), and the longest trial of each participant was selected for analysis. Two normal speed video cameras (60 fps) and three high-speed video cameras (300 fps) were used to record the landing point of the ball and upper body motion during the throw-in, respectively. In this study, due to the limited size of the indoor facility (the distance from the release point to the opposite side wall was about 25 m), the exact distance of throws more than 25 m could not be measured directly. In this case, ball distance was estimated using a quadratic regression curve fitted to the coordinates of the ball 10 frames before it collided with the wall. White markers were fixed onto several anatomical landmarks of the upper body (top of the head, acromion, radial head, styloid process of ulna, head of the third metacarpal bone, greater trochanter). The marker coordinates were smoothed by a fourth order Butterworth low-pass filter with a 20 Hz cut-off frequency. The global coordinate system was expressed as follows: the Z axis was vertical and pointed upward, the Y axis was horizontal and pointed in the throwing direction, and the X axis was perpendicular to the Z and the Y axis.

Joint coordinate systems of the trunk, shoulder, elbow, and wrist were fixed on the greater trochanter, acromion, radial head, and styloid process of the ulna, respectively (Figure 5.1). Each joint angle was expressed as follows: X_t is the axis of trunk flexion/extension, Y_t is the axis of trunk lateral bending, Z_t is the axis of trunk longitudinal rotation, X_s is the axis of shoulder horizontal adduction/abduction, Y_s is the axis of shoulder flexion/extension (adduction/abduction), Z_s is the axis of shoulder internal/external rotation, X_e is the axis of shoulder inversion/eversion, Y_e is the axis of elbow flexion/extension, Z_e is the axis of forearm pronation/supination, X_w is the axis of wrist palmar/dorsal flexion, Y_w is the axis of radial/ulnar flexion, Z_w is the axis of forearm pronation/supination. It should be noted that the Y_s axis representing shoulder flexion/extension, could partially represent the axis of shoulder adduction/abduction when the upper arm is in a horizontally abducted posture. From the procedure of Sprigings *et al.* (1994), the contribution of each joint rotational motion (angular velocity) to the hand velocity within the sagittal plane, from foot contact to ball release was computed.

The ensemble average of hand velocity at ball release, ball velocity, ball release angle, and throw-in distance of all the subjects were 12.1 ± 1.4 m/s, 15.1 ± 1.8 m/s, 31.9 ± 6.0°, and 20.3 ± 4.6 m, respectively. There was a strong correlation between the ball velocity and the throw-in distance (r = 0.90), while the release angle of the ball (which ranged from 18.8 to 40.9°) showed a weak, non-significant correlation with throw-in distance (r = 0.25).

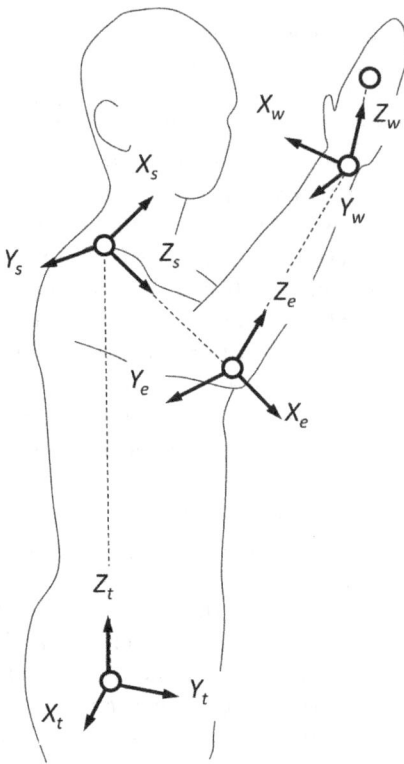

Figure 5.1 Definitions of joint coordinate systems of upper body (modified from Shinkai, 2014)

Linthorne and Everett (2006) reported that the optimum release angle of a throw-in was approximately 30°, and any slight difference from this angle would have little influence on the distance. Combining our current results and the findings of Linthorne and Everett (2006), it is reasonable to consider that throw-in distance most likely depends on initial ball velocity. In addition, the hand velocity at ball release strongly correlated with ball velocity ($r = 0.91$) and throw-in distance ($r = 0.85$). These results support the theory that enhancing the hand velocity is a very important factor for a long throw-in.

From the group of participants, we selected one typical long thrower (age 23 yrs; height 177.0 cm; body mass 77.9 kg; career 14 yrs; position full-back) who achieved an outstanding throw-in distance (30.7 m). The throw-in distance was the longest recorded in the literature and corresponds to a distance beyond the near goalpost from the touchline. The initial parameters of this participant's throw-in were: hand velocity = 14.7 m/s, ball velocity =19.0 m/s and ball release angle = 29.1°.

To identify the specific characteristics of his throwing technique, the contribution of the upper body motions to hand velocity (Figure 5.2) was

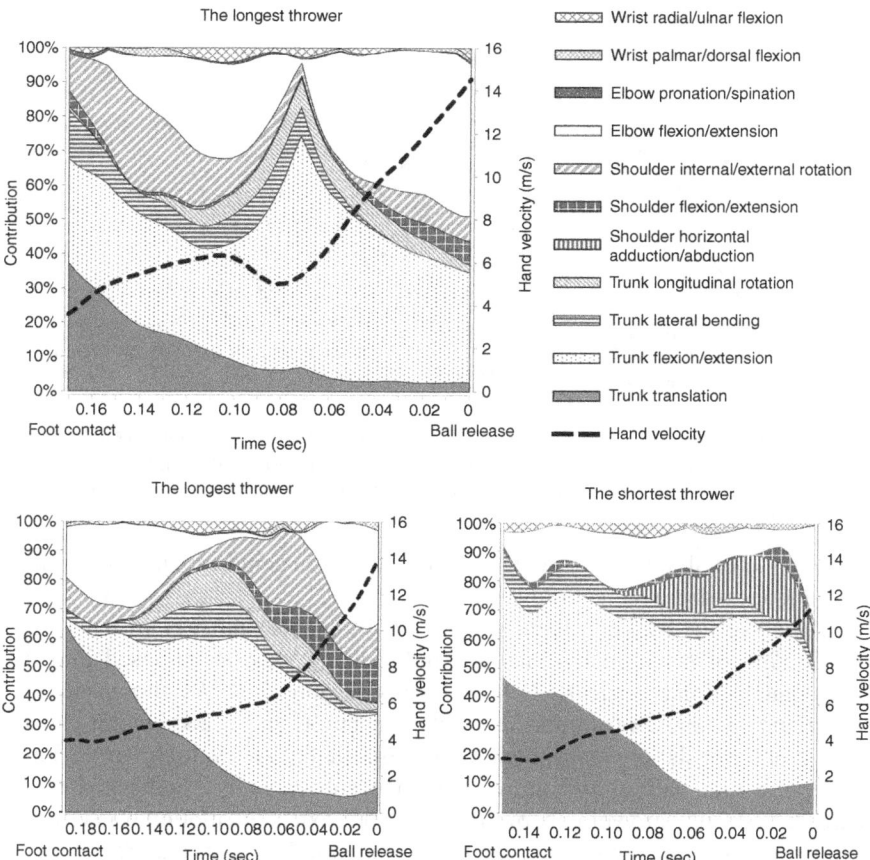

Figure 5.2 The contribution of each joint motion of the upper body to the hand velocity of the longest thrower (top), an average thrower (bottom left), and the shortest thrower (bottom right) (modified from Shinkai, 2014)

compared to that of an average thrower (22.1 m, left bottom panel of Figure 5.2) and the shortest thrower (15.1 m, right bottom panel of Figure 5.2).

As shown in Figure 5.2, the three players consistently showed that trunk translation was the main contributor to hand velocity during the first half of the movement, and trunk flexion (forward rotation) showed the largest contribution during the second half. In addition, elbow pronation/supination, wrist palmar/dorsal flexion, and wrist radial/ulnar flexion contributed minimally to hand velocity.

There were marked differences regarding the contribution of the throwing arm motions. The longest thrower exhibited a distinctively larger contribution of elbow extension, which dominated most of the contribution made by the throwing arm. In contrast, the average thrower showed notable contributions of shoulder internal rotation and extension. Moreover, the shortest thrower exhibited a unique

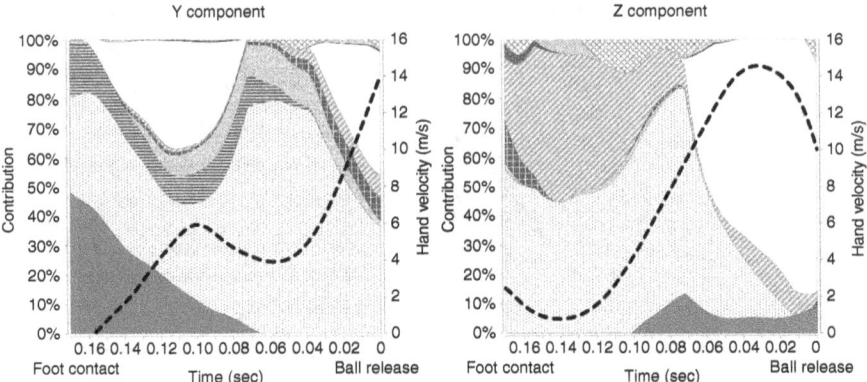

Figure 5.3 Contribution of each joint motion of the upper body to hand velocity (left panel: Y component, right panel: Z component) (modified from Shinkai, 2014). Explanatory notes for individual contributions are same as those in Figure 5.2

contribution of shoulder horizontal adduction, whose contribution was minimal for the longest and average throwers.

To analyse the motion characteristics of the longest thrower in detail, the contribution of his joint motions to hand velocity was decomposed into the horizontal (Y) and vertical (Z) components (Figure 5.3). As shown, unique transitions for the contribution of trunk translation and trunk flexion can be seen. Trunk translation contributed to increase the forward velocity of the hand during the first half of the throw, and then subsequently contributed to an increase in the upward velocity of the hand. Knee extension motion was visually confirmed from the video records and could potentially trigger that transition by changing the direction of the trunk momentum. On the other hand, trunk flexion was the main contributor for both the forward and upward velocity of the hand. However, the contribution for the upward velocity suddenly disappeared during the final phase yet remained the largest contributor to the forward velocity. The longest thrower showed a distinctively larger trunk flexion angular velocity than those of the other players (long thrower = 408.0°/s vs. others = 321.7 ± 42.3°/s). Thus, the unique transition whereby trunk flexion exclusively contributes to the forward velocity of the hand might be an efficient way to utilize the large trunk flexion angular velocity for achieving a faster forward hand velocity. Such a strategy, as used by a distinctive long thrower, seems appropriate because ball velocity is a strong determinant of throw distance rather than release angle.

In the case of the longest thrower, elbow extension had a remarkable contribution to hand velocity in both directions (forward and upward). It can be considered that this contribution of elbow extension in both directions was due to the more horizontally adducted posture of his shoulder from the pulled back position to ball release. In this study, this posture was defined as the angle between the X axis and the upper arm vector in the frontal (XY) plane, and this angle at the elbow maximum flexion of the long thrower and the others were 57.7° and 37.2 ± 12.3°

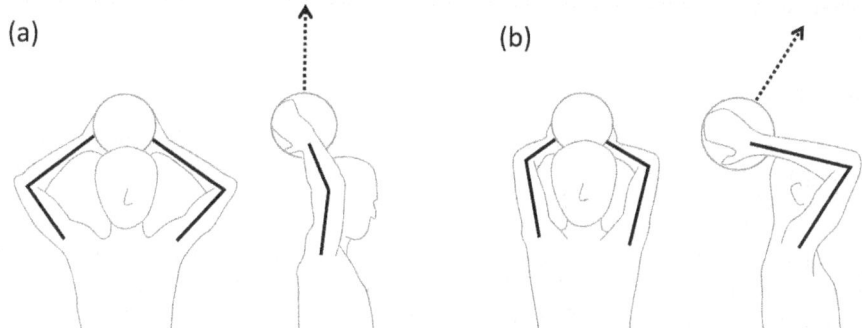

Figure 5.4 A schematic to illustrate the differing effect of elbow extension on ball velocity in two different shoulder horizontal adduction positions at the pulled back position. The shoulders of the longest thrower were more horizontally adducted (b) than the other players (a), and this posture would be work effectively to increase forward hand velocity (modified from Shinkai, 2014)

respectively. In most players, since the shoulder is not horizontally adducted (i.e. the posture of spreading both elbows) at the pulled back position, elbow extension acts to move the hand upward (Figure 5.4a). In this case, the thrower needs to utilize other joint motions such as shoulder extension and internal rotation to generate a forward velocity of the hand. In contrast, the longest thrower had a posture of tightening both elbows at the pulled back position. This posture allowed him to use elbow extension to directly increase hand velocity not only for the upward direction but also for the forward direction (Figure 5.4b). The unique elbow tightening posture brought a further advantage in that he can maximally utilize his substantially larger elbow extension angular velocity (long thrower = 1303.2°/s vs. others = 1038.7 ± 153.5°/s) for accelerating the hand in both directions.

In summary, notable characteristics of a long throw-in technique were highlighted to be quick trunk flexion and elbow extension with a specific shoulder posture, which can maximally utilize elbow extension for increasing hand velocity. It should be noted that sufficient muscle strength of the upper body might be necessary to conduct a throw-in motion in that manner. Therefore, introduction of these motions may not always instantly improve throw-in distance for all soccer players. A limitation of this study is that these characteristics of the motion were only obtained for one long thrower. Further research which accumulates biomechanical data of the long throw-in is necessary to be able to generalize the motion characteristics of this skill.

4. Conclusion

We succeeded in illustrating the characteristic upper limb motions of a long distance thrower who is capable of throwing the ball greater than 30 m. His large hand velocity at ball release was produced by a quick forward flexion of the trunk and a rapid elbow extension, while maintaining a more horizontally adducted posture of the shoulders throughout the forward swing of the arms. These main findings can be translated into useful coaching cues for technical training to lengthen throw-in distance.

References

Acar, M. F., Yapicioglu, B., Arikan, N., Yalcin, S., Ates, N., & Ergun, M. (2009). Analysis of goals scored in the 2006 World Cup. In T. Reilly & F. Korkusuz (Eds.), *Science and Football VI* (pp. 235–242), New York: Routledge.

Barraza, L. C. H., & Yeow, C. H. (2015). An investigation of full body kinematics for static and dynamic throw-in in proficient and non-proficient soccer players when they tried to hit a specific target. In F. Colloud, M. Domalain & T. Monnet (Eds.), *Proceedings of 33th International Conference of Biomechanics in Sports* (pp. 956–959), Poitiers: France.

Bray, K., & Kerwin, D. G. (2004). Modelling the long throw in soccer using aerodynamic drag and lift. In M. Hubbard, R. D. Mehta & J. M. Pallis (Eds.), *The Engineering of Sport 5* (pp. 56–62), International Sports Engineering Association.

Cerrah, A. O., Şimşek, D., & Ertan, H. (2012). The evaluation of ground reaction forces during two different soccer throw-in techniques: a preliminary study. In F. Colloud, M. Domalain & T. Monnet (Eds.), *Proceedings of 30th International Conference of Biomechanics in Sports* (pp. 299–302), Melbourne: Australia.

Chang, J. (1979). The biomechanical analysis of the selected soccer throw-in techniques. *Asian Journal of Physical Education, 2*, 254–260.

Kerwin, D. G., & Bray, K. (2004). Quantifying the trajectory of the long soccer throw-in. In M. Hubbard, R. D. Mehta & J. M. Pallis (Eds.), *The Engineering of Sport 5* (pp. 63–69), International Sports Engineering Association.

Kline, L. E., & Samonisky, M. (1981). The soccer throw-in. *Journal of Physical Education, Recreation and Dance, 52*, 57–59.

Kollath, E., & Schwirtz, A. (1988). Biomechanical analysis of the soccer throw-in. In T. Reilly, A. Lees, K, Davids & W. J. Murphy (Eds.), *Science and Football* (pp. 460–467), London: E & FN Spon.

Lees, A. (2013). Biomechanics applied to soccer skills: Soccer throw-in. In M. A. Williams (Ed.), *Science and Soccer: Developing elite performers (Third Edition)* (pp. 224–228), Oxford: Routledge.

Lees, A., Kemp, M., & Moura, F. (2005). Biomechanical analysis of the soccer throw-in with a particular focus on the upper limb motion. In T. Reilly, J. Cabri & D. Araújo (Eds.), *Science and Football V* (pp. 89–94), New York: Routledge.

Levendusky, T. A., Clinger, C. D., Miller, R. E., & Armstrong, C. W. (1985). Soccer throw-in kinematics. In J. Terauds & J. N. Barham (Eds.), *Proceedings of 3rd International Conference of Biomechanics in Sports* (pp. 258–268), Del Mar, CA: Academic Publishers.

Linthorne, N. P., & Everett, D. J. (2006). Release angle for attaining maximum distance in the soccer throw-in. *Sports Biomechanics, 5*, 243–260.

Messier, S. P., & Brody, M. A. (1986). Mechanics of translation and rotation during conventional and handspring soccer throw-ins. *International Journal of Sport Biomechanics, 2*, 301–315.

Shinkai, H. (2014). Technical factors for increasing the throwing distance in soccer throw-in (in Japanese). *Japanese Journal of Biomechanics in Sports and Exercise, 18*, 232–239.

Sprigings E., Marshall R., Elliott B., & Jennings L. (1994). A three-dimensional kinematic method for determining the effectiveness of arm segment rotations in producing racquet-head speed. *Journal of Biomechanics, 27*, 245–254.

The International Football Association Board (IFAB). (2016) *Laws of the Game* 2016/17 (p. 99).

Part II
Direct free kicks

6 Aerodynamics of modern footballs

Takeshi Asai

1. Introduction

Aerodynamics plays a prominent role in defining the flight of a football that is struck or thrown through the air. Past research on the aerodynamic characteristics of footballs focused on the traditional 32-panel ball comprised of pentagons and hexagons (Bray & Kerwin, 2003; Asai *et al.*, 2007; Goff & Carré, 2009; and Passmore *et al.*, 2012). However, the number of panels decreased from 32 to 14 (i.e. the Adidas Teamgeist II) for the official match ball of the 2008 Beijing Olympic games, and then down to 8 panels (i.e. the Adidas Jabulani) for the official match ball of the 2010 FIFA World cup held in South Africa. Furthermore, the new Adidas Brazuca football used in FIFA World cup 2014 had 6 panels, with the panels featuring a new shape and curved design. However, the aerodynamic characteristics of these new footballs with fewer than 32 panels are not well known. Therefore, in this chapter, the fundamental aerodynamic characteristics of the Brazuca and Jabulani, which are two of the latest footballs, are discussed using a wind tunnel test.

2. Wind tunnel

The basic structure of the wind tunnel consists of a drive section where flow is generated, a settling section where flow is settled, and a test section where measurements are taken, as shown in Figure 6.1. In the drive section, air is usually blown by a rotating fan; however, at times when flow velocity exceeds subsonic speed, compressed air is often used. The settling section is constructed to eliminate irregularity of the airflow by passing it through a large number of holes or a grid. The test section is the area where the object to be measured is placed, and the aerodynamic forces on the object are measured using a six-component balance. Test sections can be divided into a closed type, which is enclosed by a wall, or an open type without walls.

When objects to be measured can easily be removed from the open type test section and the flow is not affected by the walls, the impact of external air needs to be considered. Conversely, the closed type test section is not affected by external air but the impact of boundary layers along the walls must be taken into

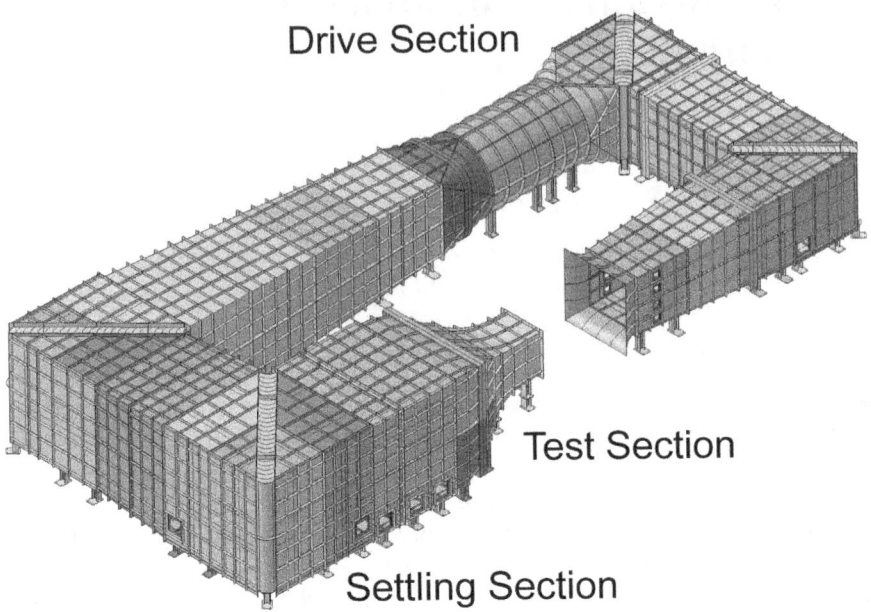

Figure 6.1 Structure of a wind tunnel

account. Furthermore, there are various methods for connecting the objects to be measured to the measuring instrument including supporting a test object from below by using a post or suspending a test object from a wire. However, it is difficult to completely eliminate the impact of the connection site on the aerodynamic forces experienced by a test object, thus posing a problem in data gathering and analysis. Recently, magnetic suspension and balance systems have been studied, where the measured object is made to float in air by using magnetism without a support system. This system is expected to be further developed and more widely used.

3. Drag of the Brazuca and Jabulani footballs

In this section, a wind tunnel test is used to compare drag characteristics for the Brazuca and Jabulani footballs. Additionally, the flight characteristics of the footballs were studied using 2D flight trajectory simulation. For the closed-circuit, low-speed, low-turbulence tests, the wind tunnel (San Technologies Co., LTD), that is mentioned above, at the University of Tsukuba was used. The wind tunnel has the following characteristics: maximum wind velocity of 55 m/s; outlet dimensions of 1.5 m × 1.5 m; wind velocity distribution within ± 0.5%; and turbulence of 0.1% or lower.

Tests were performed by placing an Adidas 6-panel Brazuca football, i.e. the official ball for the Brazil world cup, and an Adidas, 8-panel Jabulani football, i.e. the official ball for the South Africa world cup, in the wind tunnel with horizontal beam support from the back as shown in Figure 6.2. The forces acting on the footballs were measured using a sting-type, six-component force detector (LMC-61256, Nissho Electric Works). Drag coefficient (*Cd*) and lift coefficient (*Cl*) were obtained from the measured drag force (*D*) and lift force (*L*) as shown in Equations (1) and (2) below.

$$Cd = \frac{D}{\frac{1}{2}\rho U^2 A} \tag{1}$$

$$Cl = \frac{L}{\frac{1}{2}\rho U^2 A} \tag{2}$$

where ρ is the air density (1.2 kg/m³), U is the flow rate, and A is the projected area of the football (A = 0.038 m²).

For the wind tunnel tests, the critical Reynolds number for the Brazuca was approximately 2.7×10^5 at 17 m/s while the critical Reynolds number for the Jabulani was approximately 3.8×10^5 at 24 m/s. A comparison of drag

Figure 6.2 The set up for the wind tunnel experiment

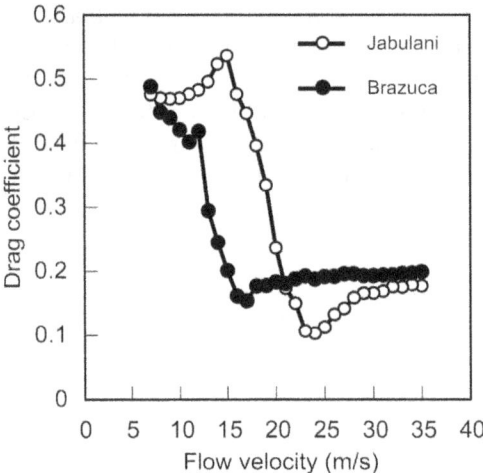

Figure 6.3 Drag characteristics of the Brazuca and Jabulani footballs. The Brazuca has lower air resistance in the intermediate velocity range ($15 < U < 22$ m/s), but the Jabulani has slightly lower air resistance in the high velocity range ($22 < U < 35$ m/s)

coefficients demonstrated a higher drag coefficient (i.e. resistance) for the Jabulani than for the Brazuca in the intermediate-speed range ($15 < U < 22$ m/s) as shown in Figure 6.3. However, in the high-speed range ($22 < U < 35$ m/s) the Brazuca displayed a higher drag coefficient. This demonstrates that for the intermediate-speed range, the Brazuca experiences less air resistance than the Jabulani. However, for the high-speed range, air resistance for the Brazuca is higher compared to that for the Jabulani.

A comparison of 2D flight trajectories for the Brazuca and the Jabulani, considering only drag coefficients (Goff *et al.*, 2010), indicates that when the ball is kicked at an initial velocity of 30 m/s and an angle of attack of 10°, the velocity of the Brazuca at a distance of 20 m is 24.22 m/s and 25.23 m/s for the Jabulani. The result demonstrates that the Brazuca is slower than the Jabulani in these conditions. Furthermore, when the balls are kicked at an initial velocity of 20 m/s and an angle of attack of 22°, the speeds at a distance of 20 m are 16.03 m/s for the Brazuca and 13.07 m/s for the Jabulani. In contrast to the high velocity condition, the Brazuca is faster than the Jabulani. Therefore, the Brazuca experiences less air resistance than the Jabulani in the intermediate-speed range, making it better suited for ball speeds in that range. Additionally, the difference in air resistance between the Brazuca and the Jabulani for the high-speed range is smaller than in the intermediate-speed range.

Footballs predating the Jabulani were tested in the wind tunnel measurements for the critical Reynolds number (Asai & Seo, 2013). The critical Reynolds

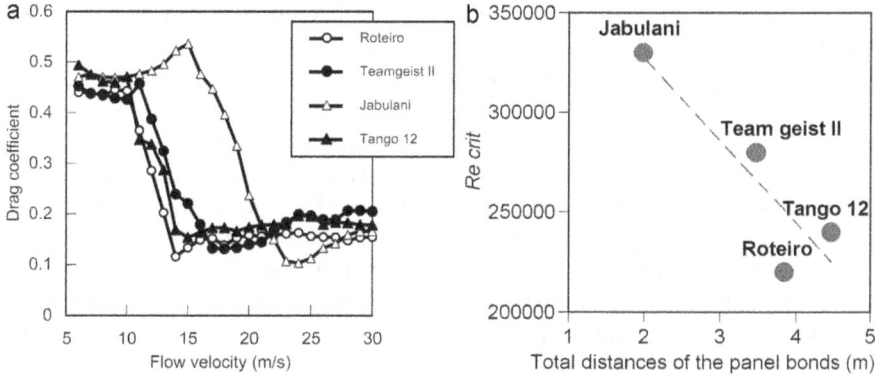

Figure 6.4 Drag coefficient (Cd) of the Roteiro, Teamgeist II, Jabulani, and Tango 12 footballs. The critical Reynolds numbers were ~2.2 × 105 (Cd ≈ 0.12), ~2.8 × 105 (Cd ≈ 0.13), ~3.3 × 105 (Cd ≈ 0.11), and ~2.4 × 105 (Cd ≈ 0.15), respectively (a), and correlation between the extended total distances of the panel bonds and the critical Reynolds number (r = 0.9) (b)

number of other types of balls: the Roteiro, the Teamgeist II, the modified Jabulani (based on a different panel position) and the Tango were approximately 2.2×10^5 for a $C_d \approx 0.12$, 2.8×10^5 for a $C_d \approx 0.13$, 3.3×10^5 for a $C_d \approx 0.13$ and 2.4×10^5 for a $C_d \approx 0.15$, respectively. The critical Reynolds number obtained for the Roteiro was the same as that reported by (Asai *et al.*, 2007). The standard C_d values for the Tango 12 and the Jabulani in the supercritical regime were approximately 0.18 and 0.15, respectively. The average C_d in the subcritical regime was approximately 0.47, which was slightly greater than that of the Jabulani at 0.44. The newer balls showed an increased critical Reynolds number while the C_d curve shifted to the right. However, the C_d curve of the Tango 12 was similar to that of the Roteiro to a greater degree than when compared to the Jabulani (Figure 6.4).

The extended total distances of the panel bonds and the number of ball panels were as follows: Adidas Roteiro: 3840 mm, smooth surface with 32 pentagonal and hexagonal panels; Adidas Teamgeist II: 3470 mm, small protuberance with 14 panels; Adidas Jabulani: 1980 mm, small ridges or protrusions with 8 panels; and Adidas Tango 12: 4470 mm, small grip texture with 32 panels. A high degree of correlation was observed between the extended total distances of the panel bonds and the critical Reynolds number (r = 0.9) as indicated in Figure 6.4b. The critical Reynolds number of each ball decreases with the number of panels, i.e. it decreases from the Roteiro to the Teamgeist II to the Jabulani. Furthermore, the extended total distance of the panel bonds decrease based on the number of panels. The Tango 12 ball has 32 panels; therefore, its extended total distance of panel bonds will increase, and its critical Reynolds number will be similar to that of the 32-panel Roteiro ball.

Achenbach (1974) reported that an increase in the roughness of the spherical surface decreases the critical Reynolds number. Therefore, it can be concluded that the roughness of the ball surface increases with the extended total distance of panel bonds, thereby causing the critical Reynolds number to decrease. In terms of roughness, the panel surface of the Roteiro is relatively smooth; the Teamgeist II has small protuberances; the Jabulani has small ridges; and the Tango 12 has small grip textures. The critical Reynolds number of the Roteiro was lower than that of the Jabulani despite the panel surface of the Roteiro being relatively smoother than that of the Jabulani. The small designs and appliques applied to the football panels by the manufacturer appeared to play only a minor role in this experiment. Therefore, the critical Reynolds number of a football, as considered within the scope of this experiment, may depend on the extended total distance of the panel bonds.

4. Drag and lift of a spinning football

Drag and lift forces of a spinning football display properties that are different to when they do not spin but are instead fixed in a wind tunnel (Watts & Ferrer, 1987; Asai *et al.*, 2007). Although the separation point for the boundary layer varies slightly for non-spinning and low-spin footballs, on average it is positioned almost symmetrically to the direction of travel. Additionally, there is little deflection in the wake as can be seen in Figure 6.5a. In contrast, for spinning footballs, the boundary-layer separation point is asymmetrical because of the effect of rotation that also results in corresponding wake deflection. This wake deflection is one of the factors that generate lift (Figure 6.5b). When compared with the position of the boundary-layer separation point for a non-spinning ball, the separation point on the side rotating in the same direction as the dominant flow often moves backwards for a spinning ball while the separation point on the side rotating in the opposite direction often moves forward.

Drag crisis, as seen for non-spinning balls, is not observed for drag of a spinning football. The tendency for drag to increase proportionally because of increased spin parameters (*Sp*) is demonstrated in Figure 6.6a and is indicated by Equation 3 below. Lift is generated because of the Magnus effect (Magnus, 1953), and the tendency for this lift to increase in line with an increased *Sp* is demonstrated in Figure 6.6b. These results show that when a ball is flying, the number of ball rotations does not decrease significantly (i.e. acting rotational resistance is low). However, the speed of the ball reduces because of drag, resulting in an increased *Sp*. Accordingly, the trajectory of a spinning ball at high speed is thought to exhibit a notably sharper curve in the latter half of the flight than in the first half (Dupeux *et al.*, 2011).

$$Sp = \frac{d\omega}{U} \tag{3}$$

Here, d is the ball's diameter, U is the flow rate, and ω is the angular velocity.

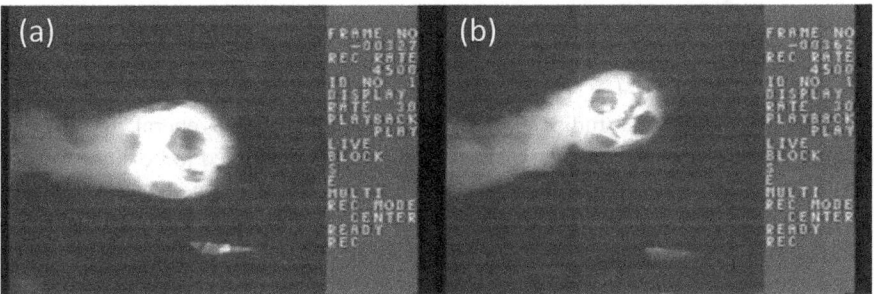

Figure 6.5 Flow visualisation by using a smoke agent for: (a) low spinning football; and (b) spinning football. (a) Top-view image of low spinning football (27 m/s; 0.5 RPS, anti-clockwise) shows the separation point was ~120° from the front-stagnation point. (b) Top-view image of spinning football (27 m/s; 7 RPS, anti-clockwise) shows the vortex deflection and downwash due to the effects of a relative difference in fluid speed caused by the spinning

Furthermore, there is a possibility that at low spin, the reverse Magnus effect can be observed (Briggs, 1959). The reverse Magnus effect occurs when a force emerges in the direction opposite to that of the regular Magnus force. This situation occurs because of the lift that is generated based on the timing when the boundary layer changes from a laminar boundary layer to a turbulent boundary layer, which varies depending on the surface of the ball (Taneda, 1957). However, few reports exist on the reverse Magnus effect in relation to footballs and the details are an issue for the future.

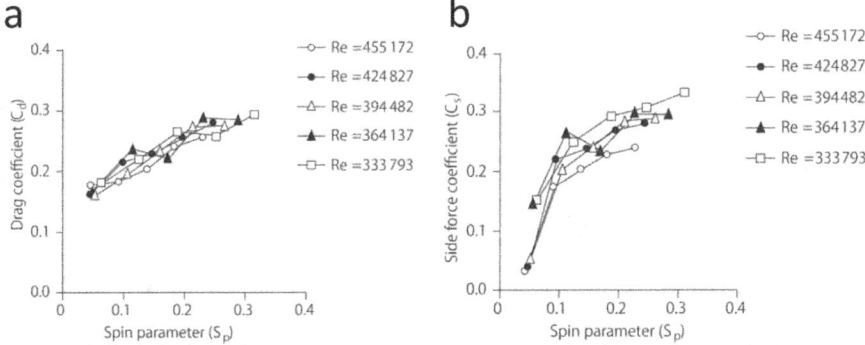

Figure 6.6 Relationship between *Sp, Cd* and *Cl* for rotating balls. The *Cd* tended to increase as the *Sp* increased. The *Cd* was largely independent of speed, but appeared to be highly dependent upon the *Sp*. The *Cl* (*Cs*) tended to increase in a curved manner as the *Sp* increased

5. Lift of Brazuca and Jabulani footballs

In this section, a wind tunnel test was used to compare the lift characteristics for the Brazuca and the Jabulani for non-spinning conditions and when using an impact-type kicking robot. The impact of these lift characteristics on actual football flight trajectories was studied by Hong & Asai (2014). In general, the average stationary lift force of spheres, such as footballs, is considered to be close to zero; however, non-stationary lift forces often exhibit irregular fluctuations. Using a wind tunnel, the mean deviation for lift force fluctuation for the Brazuca was 0.6 N and for the Jabulani was 1.7 N at a wind velocity of 20 m/s. The lift force fluctuation tended to be smaller for the Brazuca than for the Jabulani (Figure 6.7a, 6.7b). Furthermore, when using an impact-type kicking robot, the average deviation of the landing point in a goal located at a distance of 25 m was larger for the Jabulani

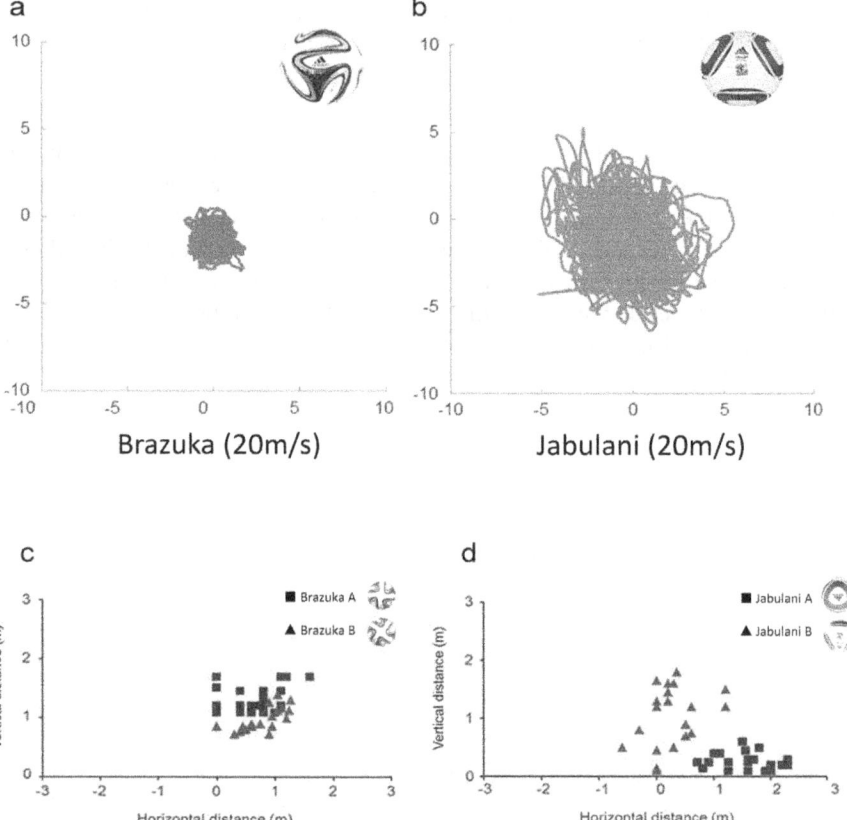

Figure 6.7 Scatter plots of the side and lift forces of Brazuca (a) and Jabulani (b), and comparison of the flight characteristics (i.e. points of impact) of the different balls for different panel orientations (i.e. initial launch velocity of 30 m/s and an angle of 15°) of Brazuca (c) and Jabulani (d)

(0.39 m) than for the Brazuca (0.27 m). Therefore, it can be surmised that lift fluctuations in free flight are smaller for the Brazuca than for the Jabulani. The reason for this lift fluctuation for non-spin shots is presumed to be boundary layer instability (Asai & Kamemoto, 2011) as well as the effect of surface design including the panels, position, length, and depth of the seams of the ball.

A wind tunnel was used to perform boundary layer visualisation through Particle Image Velocimetry (PIV) in order to study the impact of ball panels and seam position on boundary layer instability. The 2D-PIV measurements were

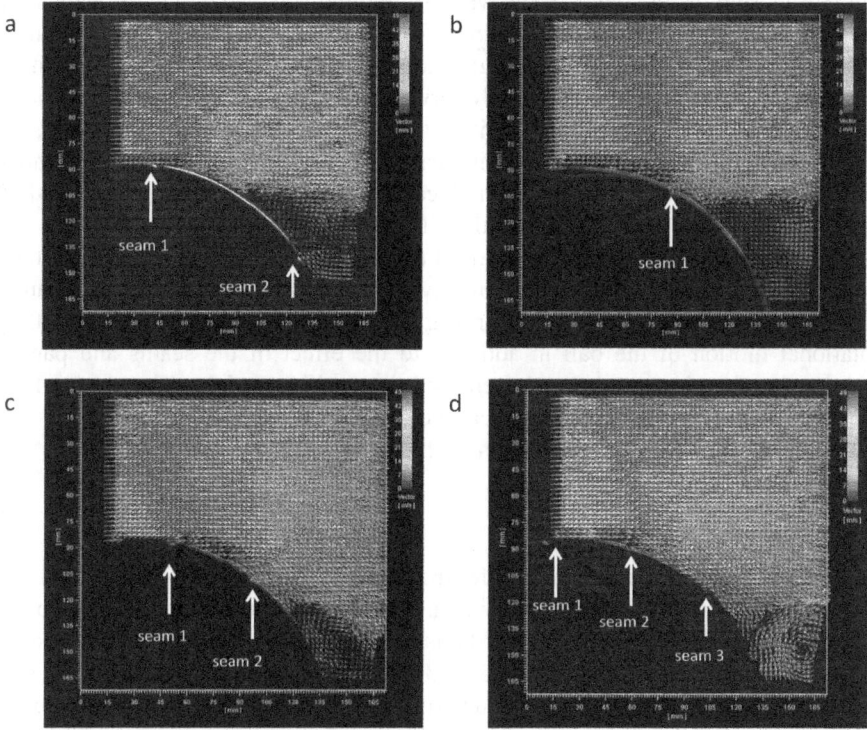

Figure 6.8 Velocity vectors on the suction side of the footballs at $U = 30$ m/s. (a) In the case of two seams spaced approximately 80 mm apart, such that the angle formed with the stagnation point in seam 1 was 95° and with that in seam 2 was 150°; the separation point was found to be approximately 120°, nearly 65 mm from the centre of the ball. (b) For the orientation with a single seam (i.e. 120° in seam 1), the separation point was approximately 125°, approximately 70 mm from the centre of the ball and shifted slightly backwards. (c) The seams affected the flow of the air and the separation point was moved to approximately 140° (approximately 85 mm), along the face with two seams with relatively narrow spacing of approximately 50 mm between them, (i.e. 100° in seam 1, and 130° in seam 2). (d) For the face with three seams, and the spacing between them being approximately 45 mm, (80° in seam 1, 105° in seam 2, and 130° in seam 3), the air flow around the football re-attached at seams 2 and 3, and the separation point was at approximately 145° (approximately 90 mm), having moved to the rearmost position

carried out on the centreline of the football. Micro-droplets with diameters of 1 μm were generated by an aerosol generator (PivPart40, PivTec), and were introduced into the flow by a sirocco fan into the wind tunnel. A high-repetition, pulsed ND:Yag laser (LDP-100MQG, Lee Laser) was used to illuminate the micro-droplet particles. A high-speed camera (Memrecam GX-8, Nac) was used to record the images at a sampling frequency of 1000 Hz. The airflow velocity was set to 30 m/s.

From the tests, it was determined the boundary-layer separation point does not remain in one place on the ball surface. Instead, there is a tendency toward instability. Accordingly, the range of the wake behind the ball also fluctuates, which is thought to be one of the factors behind lift fluctuations. Furthermore, the seams of a ball become significant with regard to the separation point and lift fluctuation. If the seams were moved slightly forward in relation to the usual separation site (i.e. into the oncoming wind), it is thought that if separation occurred, it would reattach further downstream indicating the impact of the seam is minor (Figure 6.8a). However, if the seam is located near the usual separation site, the seam is thought to excite the separation (Figure 6.8b). Moreover, in the case of contiguous seams, separation is surmised to be affected by the interaction of the vortices generated by the seams themselves as can be seen in Figure 6.8c and 6.8d. The boundary layer of a ball in flight is often subject to the effect of the rotational motion of the ball in addition to the effect of the seams and panel shapes; thereby, demonstrating extremely complex fluctuations. These complex 3D fluctuations of boundary layers can be considered to be a significant reason behind lift fluctuations of footballs in flight.

6. Conclusion

The aerodynamic characteristics of footballs, although essentially equivalent to the characteristics of a sphere, vary significantly because of the effect of ball panel design and seam location. For balls in actual flight, the effect of rotation, flow of natural air, starting vortices, and subsequent vortex growth cannot be ignored. The trajectory of the ball is determined through the intricate interactions of these factors. When compared to balls used for other sports, footballs are launched onto the market in very fast cycles and tend to be used as official match balls. Therefore, the potential for developing new technologies and tactics by ascertaining and utilising these aerodynamic characteristics should be a significant area of interest.

References

Achenbach, E. (1974). Vortex shedding from sphere. *Journal of Fluid Mechanics, 62*, 209–221.

Asai, T., Seo, K., Kobayashi, O., & Sakashita, R. (2007). Fundamental aerodynamics of the soccer ball. *Sports Engineering, 10*, 101–109.

Asai, T., & Kamemoto, K. (2011). Flow structure of knuckling effect in footballs. *Journal of Fluid Structures*, *27*, 727–733.

Asai, T., & Seo, K. (2013). Aerodynamic drag of modern soccer balls. *SpringerPlus*, *2*, 171.

Bray, K., & Kerwin, D. G. (2003). Modelling the flight of a soccer ball in a direct free kick. *Journal of Sports Sciences*, *21*, 75–85.

Briggs, L. J. (1959). Effect of spin and speed on the lateral deflection (curve) of a baseball; and the Magnus Effect for smooth spheres. *American Journal of Physics*, *27*, 589–596.

Dupeux, G., Cohen, C., Goff, A. L. Quéré D., & Clanet C. (2011). Football curves. *Journal of Fluid Structures*, *27*, 659–667.

Goff, J. E. & Carré, M. J. (2009). Trajectory analysis of a soccer ball. *American Journal of Physics*, *77*, 1020–1027.

Goff, J. E., Asai, T., & Hong, S. (2010). A comparison of Jabulani and Brazuca non-spin aerodynamics. *Proceedings of the Institution of Mechanical Engineers, Part P: Journal of Sports Engineering and Technology*, *228*, 188–194.

Hong, S. & Asai, T. (2014). Effect of panel shape of soccer ball on its flight characteristics. *Scientific Reports*, *4*, 5068.

Magnus, G. (1953). *Poggendorf's Annual on Physics and Chemistry*, 88.

Passmore, M., Rogers, D., Tuplin, S., Harland, A. R., Lucas, T., & Holmes, C. (2012). The aerodynamic performance of a range of FIFA-approved footballs. *Proceedings of the Institution of Mechanical Engineers, Part P: Journal of Sports Engineering and Technology*, *226*, 61–70.

Taneda, S. (1957). Negative Magnus Effect. *Reports of Research, Institute for Applied Mechanics*, 123–128.

Watts, R. G. & Ferrer, R. (1987). The lateral force on a spinning sphere. *American Journal of Physics*, *55*(1), 40–44.

7 Impact phase of the knuckle shot

Sungchan Hong and Takeshi Asai

1. Introduction

Kicking technique for skillful ball manipulation in passes and shots is extremely important in soccer (Alcock *et al.*, 2012; Lees *et al.*, 2010; Nunome *et al.*, 2006). The outcome of a game is often determined by the ability of an athlete to kick a ball with an unpredictable trajectory. For this reason, quantitative measurement and analysis of kicking technique is necessary for coaches to understand the learning and implications of these skills (Asami & Nolte, 1983; Asai *et al.*, 2004; De Witt & Hinrichs, 2012; Dörge *et al.*, 2002; Juárez *et al.*, 2011; Katis & Kellis, 2010; Levanon & Dapena, 1998; Luhtanen, 1988; Naito *et al.*, 2010; Nunome *et al.*, 2002). Many different kicking techniques exist, all of which have a very similar impact time: about 1/100th of a second for the impact between the foot and the ball (Nunome *et al.*, 2006). This impact time maybe short, but it controls the direction of the ball after it is hit. However, the complex mechanical interactions between ball and foot are not currently fully understood to aid us in informing the coaching process of this skill.

In recent years, a commonly observed kicking technique is non-rotating non-spinning shot in which the trajectory of the ball experiences a sudden drop or swing – the so-called knuckle shot, or the knuckle effect (Asai, *et al.*, 2008). In this type of shot, the ball speed rapidly increases from 0 m/s to 20–30 m/s at 1/100th of a second and decreases thereafter due to air resistance, while the trajectory changes irregularly (Hong *et al.*, 2010). Also, knuckle shots make it more difficult for a goalkeeper to save the shot. Very few studies so far have investigated the influence of kicking technique or the ball's impact characteristics on the success of a knuckle shot (Hong & Asai, 2011). Possible reasons for this include the difficulty in evaluating the phenomenon as it occurs at extremely high speed. Moreover, there are relatively few soccer players who can intentionally execute a knuckle shot, and the success rate is low among those who can (Hong & Asai, 2011; Hong *et al.*, 2011, 2012).

In this chapter, a high-speed video camera system was used to investigate player technique during a knuckle shot. University soccer players who are capable of kicking the knuckle shot were selected, and the swing motion of their kicking leg were compared as they kicked a knuckle, curve, or straight shot. This chapter describes the impact phase and other technical aspects of the knuckle shot in soccer. Thus, in this chapter, we can explain the kicking technique of a knuckle ball by analyzing leg motion and the subsequent phase of impact between the foot and ball. A scientific clarification of the effects of different styles of kick in soccer would facilitate the learning of a more efficient shooting technique, contribute to enhanced player performance, and improve physical education teaching methods in schools.

2. Method

Two high-speed cameras (Fastcam; Photron Inc., Tokyo, Japan; 1000 fps, 1024 × 1024 pixels) were used to record the overhead and the side view of ball impact. In addition, two semi-high-speed cameras (EX-F1; Casio Computer Co. Ltd., Tokyo, Japan; 300 fps; 720 × 480 pixels) were positioned at the rear, and on the kicking leg sides (Figure 7.1).

A calibration frame (2.0 × 2.0 × 2.0 m) with 27 control points was used to calibrate the performance area. A digitizing system (Frame DIAS IV; DKH Inc., Tokyo, Japan) was employed to manually digitize the anatomical body landmarks. The direct linear transformation (DLT) method (Abdel-Aziz & Karara, 1971) was used to obtain three-dimensional (3D) coordinates of each landmark. The performance area was calibrated with a net root mean square error of 5 mm. The 3D coordinates were expressed as a right-handed orthogonal reference frame fixed on the ground (X-axis: the horizontal forward direction from the ball location; Y-axis: the horizontal left direction; Z-axis: the vertical direction).

From the images taken by the two semi-high-speed cameras (300 fps), two-dimensional (2D) coordinates were obtained through digitizing. After digital filtering with a three-point moving average of the 3D coordinates, the velocity of each joint in the kicking leg during the shot, ball velocity, and ball angular

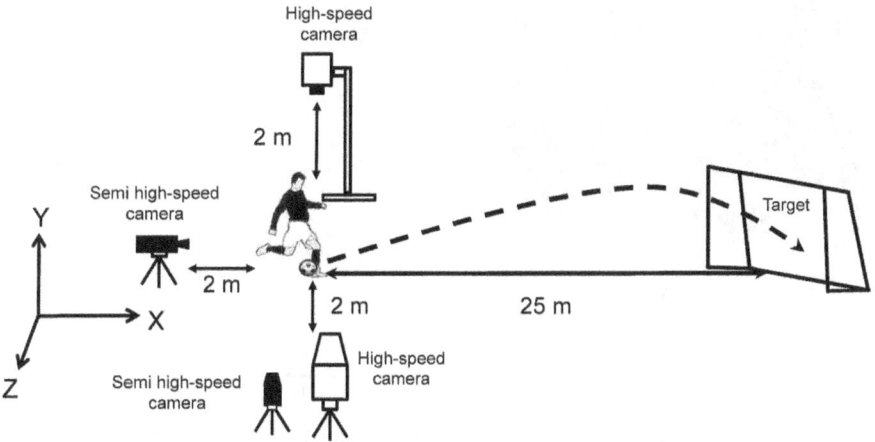

Figure 7.1 Diagram of experimental setup

velocity were determined, in addition to the joint torques at the kicking leg during the swing phase.

From the high-resolution images obtained at 1000 fps, ankle joint displacement, velocity, and angle of attack between the face and swing vectors were calculated for the impact process. Ankle joint and ball kinematics were also analyzed.

The face vector at ball impact was defined as a normal vector to the longitudinal axis of the foot (Miura & Sato, 1998). The swing vector at ball impact was defined as the velocity vector of the midpoint between the toe and the heel. The angle of attack was defined as the angle between the face and swing vectors.

Data were analyzed statistically by using analysis of variance followed by paired t-tests for multiple comparisons between variables. Variables analyzed between the three testing sessions were joint torque and attacking angle. All the statistical procedures were performed using statistical software (SPSS Japan, Inc., Tokyo, Japan). Statistical significance was set as $p < 0.05$.

3. Result & discussion

3.1 Swing phase of the kicking leg

The kicking leg motion during the swing phase of a straight, knuckle, and curve shot is shown in Figure 7.2. The side view does not show any apparent differences overall, yet the ankle joint in the curve and knuckle shots has greater dorsi-flexion than in the straight shot. In the transverse (top) and frontal (back) views, it can be seen that the curve shot has a greater swing plane angle than that of the straight shot and knuckle shot. Consequently, it can be suggested that the curve shot has a greater lateral leg swing motion than those of the straight shot, while the knuckle shot has a leg swing plane motion that is closer to that of the straight shot.

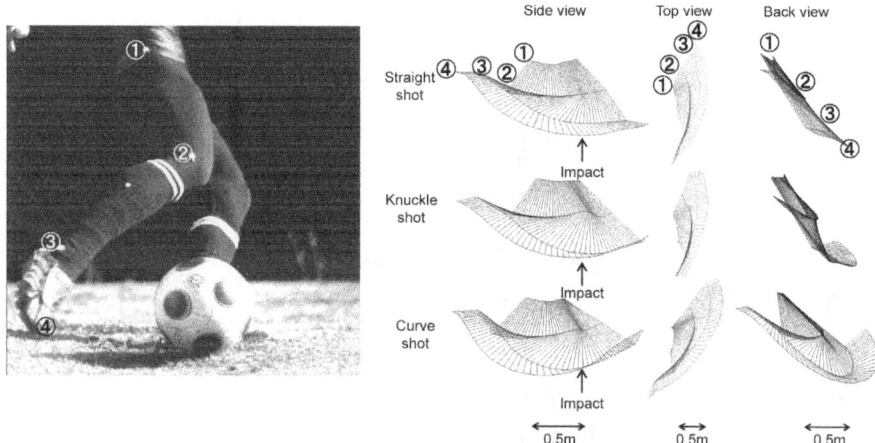

Figure 7.2 Stick diagrams of the kicking leg during the forward swing phase of a straight (top), knuckle (middle), and curve (bottom) shots

Figure 7.3 shows the forward and lateral components of linear velocity of ankle, knee, and hip joints of the kicking leg during the straight shot, knuckle shot, and curve shot. Overall, the forward component of linear velocities for each joint exhibited a consistent trend in which the peak hip velocity comes first, followed by peak knee joint velocity, and finally the peak ankle velocity occurred just before ball impact (panels a, c and e). The mechanism, in which a number of linearly linked joints function in a chain-like relationship, transmitting energy from base to tip (flail-like action), is evident with these kicking techniques (Dyson, 1970; Robertson & Mosher, 1985).

For the lateral component (panel b, d and f), ankle velocity at ball impact was 4.1 m/s for the curve shot (panel f), 0.7 m/s for the straight shot (panel b), and 0.5 m/s for the knuckle shot (panel d). The peak ankle velocity was 9.1 m/s for the

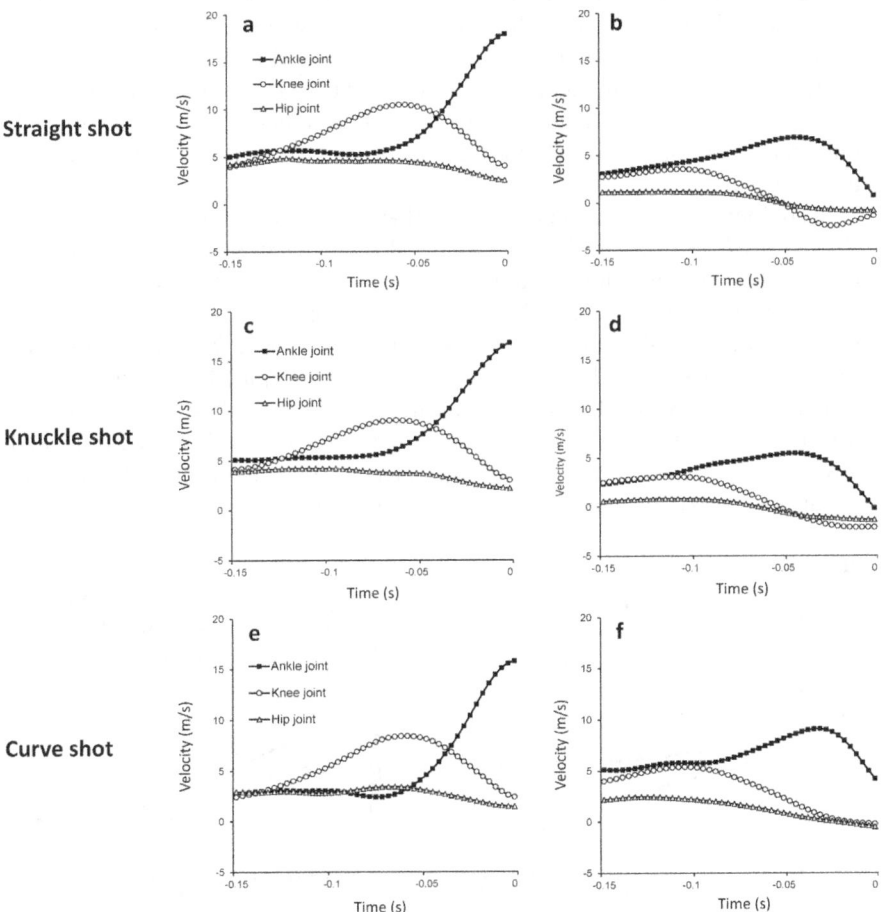

Figure 7.3 Forward and lateral components of linear velocity of ankle, knee, and hip joints of the kicking leg during the straight shot (a, b), knuckle shot (c, d), and curve shot (e, f)

curve shot, 6.8 m/s for the straight shot, and 5.4 m/s for the knuckle shot. The peak ankle lateral velocity was substantially higher in the curve shot than those of the other shots, suggesting the curved shot has a greater lateral motion. This is a key characteristic of the curve shot, which is in contrast to smaller lateral motion observed for the knuckle shot. These characteristics seen in the lateral motions can differentiate a knuckle shot from a curved shot.

3.2 Ankle velocity, ball velocity and ball rotational frequency

Figure 7.4 depicts the lateral and horizontal velocities of the ankle joint during ball impact obtained from high-speed video records. For the lateral component, the curve shot demonstrated an apparently greater reduction (deceleration) while the changes for knuckle shot and the straight shot were minimal. For the horizontal component, the straight shot exhibited a distinctively greater reduction than those of the other shots. It can be assumed that the knuckle shot approximates the straight shot in terms of lateral ankle velocity but the curve shot in terms of horizontal ankle velocity during ball impact. These results highlighted that the ankle joint motion during ball impact for the knuckle shot has a more translational nature than those of the other shots.

In addition, the maximum ankle joint velocity was 18.5 ± 0.6 m/s for the straight shot, 17.9 ± 0.7 m/s for the knuckle shot, and 17.7 ± 0.4 m/s for the curve shot. The resultant ball velocities for the straight shot, knuckle shot, and curve shot were 28.7 ± 0.9 m/s, 27.3 ± 0.9 m/s, and 27.1 ± 0.8 m/s, respectively. The average ball rotational frequency in the horizontal axis was the highest at -3.7 ± 1.2 r/s (back spin) for the straight shot, and lowest at 0.9 ± 0.4 r/s (top spin) for the knuckle shot. Moreover, on the vertical axis, the curve shot showed the highest average rotational frequency (6.8 ± 0.9 r/s (side spin)), while the knuckle shot (0.5 ± 0.2 r/s) had the lowest one. The knuckle shot was characterized by minimized ball spin while yielding a comparable ball velocity to the other shots.

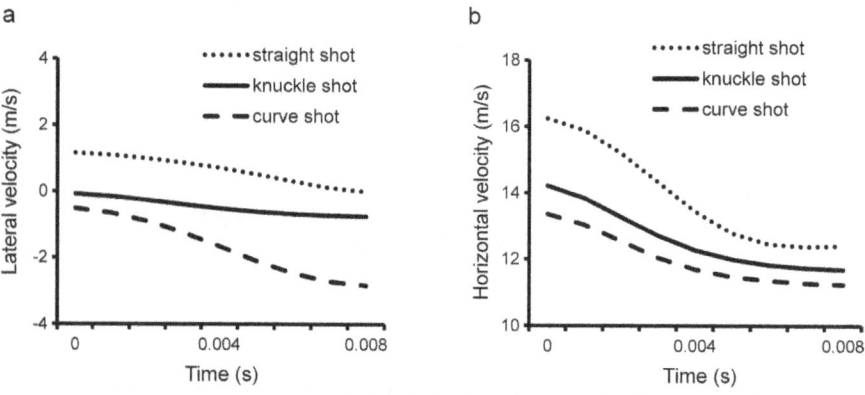

Figure 7.4 The velocity of the ankle joint during ball impact (a: lateral velocity; b: horizontal velocity)

3.3 The ankle joint during the impact phase

Figure 7.5 depicts typical ball impact images of the straight shot, knuckle shot, and curve shot. In the straight shot, we observed that the ankle joint is in a plantar flexed position (panels a and d). In contrast, the curve shots showed L-shaped dorsi-flexion of the ankle joint at ball impact (panels c and f). Furthermore, the knuckle shot (panels b and e) showed an L-shape ankle posture that is approximately similar to that of the curve shot. In other words, the ankle joint posture when impacting the ball was closer to that of the curve shot.

In Figure 7.6, the top panels (a, b and c) show typical ball impact images and the definition of the angle of attack of the three types of shots. The bottom panels (d and e) depict the average ankle displacements during ball impact during the three shots. As shown in the top panels, the average angle of attack formed by the face and swing vectors at the first frame of ball impact were $20.5° ± 3.4°$, $4.2° ± 1.5°$, and $34.6° ± 2.7°$ for the straight shot, knuckle shot, and curve shot, respectively. Significant differences were observed for the angle of attack among the three shots ($p < 0.05$). From the top view (panel d), the direction of motion in the straight shot and knuckle shot was slightly toward the inside of the swing ($11.8 ± 2.9$ mm and $3.3 ± 1.7$ mm, respectively). In contrast, in the curve shot, the direction of motion was directed toward the outside ($-45.6 ± 6.9$ mm). Significant differences were observed for the ankle displacement between the curve shot and the other shots

Figure 7.5 Representative images of the ball impact of straight (a, d), knuckle (b, e), and curve (c, f) shot (a, b, c: side views; d, e, f: top views)

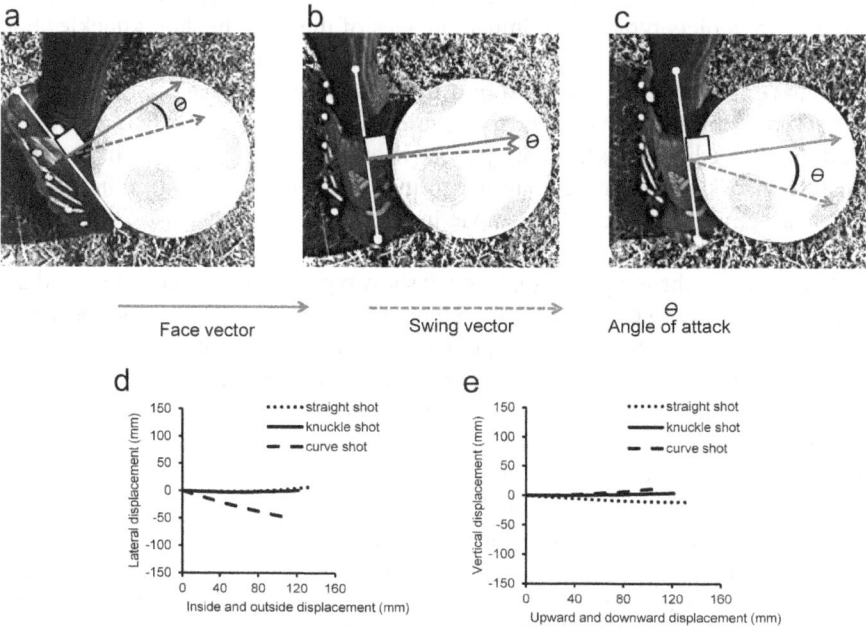

Figure 7.6 Angle of attack at ball impact of straight (a), knuckle (b), and curve (c) shot, and ankle joint displacement during ball impact for these three shots (d: top view; e: side view)

($p < 0.05$). Moreover, from the side view (panel e), the straight shot appeared to move downward (–9.6 ± 3.1 mm), whereas both the knuckle shot and curve shots moved upward (4.1 ± 1.5 mm and 13.3 ± 1.9 mm, respectively). Significant differences were observed for the vertical displacement among the three shots ($p < 0.05$).

4. Conclusions

In this chapter, we dissected the technical aspects underlying a successful knuckle shot in soccer. One of the main findings of this study was that the angle of attack in the knuckle shot (4.2° ± 1.5°) was smaller than that in the other shots (20.5° ± 3.4° and 34.6° ± 2.7° for the straight shot and curve shot, respectively). Also, the direction of motion in the knuckle shot was based on primarily translational motions (3.3 ± 1.7 mm and 4.1 ± 1.5 mm for the inside and upward motion, respectively) at the time of ball impact. A few studies have reported similar results (Hong & Asai, 2011; Hong *et al.*, 2012). This result was anticipated, considering that reduced angle of attack between the face and swing vectors at the impact surface and primarily translational motions at ball impact are fundamental technical aspects of the knuckle shot.

References

Abdel-Aziz, Y. I. & Karara, H. M. (1971). Direct linear transformation from comparator coordinates into object space coordinates in close-range photogrammetry. In *Proceedings of the ASP/UI Symposium on Close-Range Photogrammetry* (pp. 1–18), Falls Church, VA: American Society of Photogrammetry.

Alcock, A. M., Gilleard, W., Hunter, A. B., Baker, J. & Brown, B. (2012). Curve and instep kick kinematics in elite female footballers. *Journal of Sports Sciences, 30*, 384–394.

Asai, T., Akatsuka, S. J. & Haake, S. J. (1998). The physics of football, *Physics World, 11*, 25–27.

Asai, T., Takano, S., Carre, M. J. & Haake, S. J. (2004). A fundamental study of an infront curve kick in football. In M. Hubbard, R. D. Mehta, & J. M. Pallis (Eds.), *The Engineering of Sport 5* (pp. 290–295), Winfield, Kan.: International Sports Engineering Association.

Asai, T., Seo, K., Sakurai, Y., Ito, S., Koike, S. & Murakami, M. (2008). A study of knuckling effect of soccer ball. In M. Estivalet & P. Brisson (Eds.), *The Engineering of Sport 7* (pp. 555–562), Biarriz, France: Springer.

Asami, T. & Nolte, V. (1983). Analysis of powerful ball kicking. In H. Matsui, & K. Kobayashi (Eds.), *Biomechanics VIII-B* (pp. 695–700), Champaign, IL: Human Kinetics.

De Witt, J. K. & Hinrichs, R. N. (2012). Mechanical factors associated with the development of high ball velocity during an instep soccer kick. *Sports Biomechanics, 11*, 382–390.

Dörge, H. C., Anderson, T. B., Sørensen, H. & Simonsen, E. B. (2002). Biomechanical differences in soccer kicking with the preferred and the non-preferred leg. *Journal of Sports Sciences, 18*, 703–714.

Dyson, G. H. G. (1970). In *The mechanics of athletics* (5th Ed.). London: University of London Press.

Hong, S. & Asai, T. (2011). Foot and ball behaviour during impact phase of knuckling shot. *Korean Journal of Sport Science, 22*, 2330–2336.

Hong, S., Chung, C., Nakayama, M. & Asai, T. (2010). Unsteady aerodynamic force on a knuckle ball in soccer. *Procedia Engineering, 2*, 2455–2460.

Hong, S., Chung, C., Sakamoto, K. & Asai, T. (2011). Analysis of the swing motion on knuckling shot in soccer. *Procedia Engineering, 13*, 176–181.

Hong, S., Kazama, Y., Nakayama, M. & Asai, T. (2012). Ball impact dynamics of knuckling shot in soccer. *Procedia Engineering, 34*, 200–205.

Juárez, D., Mallo, J., De Subijana, C. & Navarro, E. (2011). Kinematic analysis of kicking in young top-class soccer players. *The Journal of Sports Medicine and Physical Fitness, 51*, 366–373.

Katis, A. & Kellis, E. (2010). Three-dimensional kinematics and ground reaction forces during the instep and outstep soccer kicks in pubertal players. *Journal of Sports Sciences, 28*, 1233–1241.

Lees, A., Asai, T., Andersen, T. B., Nunome, H. & Sterzing, T. (2010). The biomechanics of kicking in soccer: A review. *Journal of Sports Sciences, 28*, 805–817.

Levanon, J. & Dapena, J. (1998). Comparison of the kinematics of the full-instep and pass kicks in soccer. *Medicine and Science in Sports Exercise. 30*, 917–927.

Luhtanen, P. (1988). Kinematics and kinetics of maximal instep kicking in junior soccer players. In T. Reilly, A. Lees, K, Davids & W. J. Murphy (eds.), *Science and Football* (pp. 449–455), London: E & FN Spon.

Miura, K. & Sato, F., (1998). The initial trajectory plane after ball impact. In *Science and golf III* (pp. 535–542), Andrews, Scotland: Human Kinetics.

Naito, K., Fukui, Y. & Maruyama, T. (2010). Multijoint kinetic chain analysis of knee extension during the soccer instep kick. *Human Movement Science, 29*, 259–276.

Nunome, H., Asai, T., Ikegami, Y. & Sakurai, S. (2002). Three-dimensional kinetic analysis of side-foot and instep soccer kicks. *Medicine and Science in Sports and Exercise, 34*, 2028–2036.

Nunome, H., Lake, M., Georgakis, A. & Stergioulas, L.K. (2006). Impact phase kinematics of the instep kick in soccer. *Journal of Sports Sciences, 24*, 11–22.

Robertson, D. G. E. & Mosher, P. E. (1985). Work and power of the leg muscles in soccer kicking. In D. A. Winter, R. W. Norman, R. P. Wells, K. C. Hays & A. E. Patla (Eds.), *Biomechanics IX-B* (pp. 533–538), Champaign, IL: Human Kinetics.

8 How to kick the knuckle shot

Ball impact characteristics of top professional players and its application for training

Hironari Shinkai and Neal Smith

1. Introduction

The "knuckle ball" is characterised by a non- or low-ball spinning flight path, and its strange trajectory that moves right and left and/or up and down is like a "flying butterfly". This miracle ball has been seen in baseball pitching and volleyball serving for many years. In soccer, this type of shot began to appear frequently from around 2006 in the FIFA World Cup in Germany. The new type of ball which had a smaller number of panels connected together with thermal bonding, compared to conventional ball with 32 stitched panels, was first used in this tournament. Since then, the knuckle shot has attracted worldwide attention as a strong weapon for scoring past top-level goalkeepers.

To date, many researchers have focused on the aerodynamic characteristics of a non-spinning ball, and have revealed the reasons of this irregular flying trajectory (Hong *et al.*, 2010; Asai & Kamemoto, 2011; and many others). On the other hand, there are only a few studies that have focused on how to kick the ball to produce a non-spinning ball (Shinkai & Nunome, 2008; Shinkai, 2010; Hong *et al.*, 2013; Nunome *et al.*, 2014). It is assumed that the main reason of this small body of literature is the technical limitations of the players: there are few players who can kick a successful knuckle shot with high consistency in an actual game.

The purpose of this chapter is to illustrate the unique ball impact characteristics of a knuckle shot using examples of two top international footballers, and to verify the effect coaching has on performing the knuckle shot, derived from ball impact characteristics.

2. Ball impact characteristics of Japanese top player Keisuke Honda

Keisuke Honda (age at the time 21yrs; height 1.82m; body mass 76.4kg; shoe size US 9.5) plays for AC MILAN in Italy, and the Japan national squad. He is also well known as the leading Japanese exponent of a knuckle shot since he became professional. To aid in the explanation of how his skill is performed, we conducted a kinematic analysis of his knuckle shot, and tried to reveal how to hit the ball to produce a non- or low-spinning ball (Shinkai & Nunome, 2008).

The experiment was conducted outdoors on a natural grass surface with windless or light wind conditions. The distance between the ball and the goal was 25m to simulate a direct free kick in front of goal. The ball used was the Team Geist+ (Adidas Japan K.K., inflation: 1000 hPa) which was used in the FIFA World Cup 2006. Two normal speed video cameras (60 fps) and two high-speed video cameras (250 fps) were used for recording the ball flying trajectory and launch velocity and angle, respectively. Ten maximal knuckle shots were performed, and two ultra-high-speed video cameras (2000 fps) were used to capture the shank, foot and ball motion during the ball impact phase. White markers were fixed onto several anatomical landmarks of the kicking limb (head of fibula, lateral malleolus, lateral side of calcaneus, 5th metatarsal base, and 5th metatarsal head). The Global coordinate system was expressed fixed on the ground: Z axis was vertical and pointed upward, Y axis was horizontal and pointed to the goal, and X axis was perpendicular to the Z and the Y axis. Foot velocity and initial ball velocity were measured from the coordinates of the 5th metatarsal base just before ball contact and centre of the ball just after ball contact, respectively. The number of ball rotations were obtained from the lateral side image (2000 fps) by applying the method of Jinji and Sakurai (2006). Tri-axial angular displacements of the ankle (plantar/dorsal flexion, abduction/adduction, inversion/eversion) during ball contact were measured by the method of Shinkai *et al.* (2009). Shank and foot inclination angles at initial ball contact were defined as the angle between X axis and shank segment vector (S_{shank}) and foot segment vector (S_{foot}) in the frontal (XZ) plane, respectively (Figure 8.1a). An index representing the foot position relative to foot swing velocity (angle of attack) was obtained as IP_{ML}. This was defined as the angle between the velocity vector of the foot (V_F) and the vector (R_F) defined by the vector product of the vector pointing from the lateral side of calcaneus toward lateral malleolus and S_{foot}, perpendicular to S_{foot} (Figure 8.1b). Another index of the differences in foot contact point along its longitudinal axis (IP_{LA})

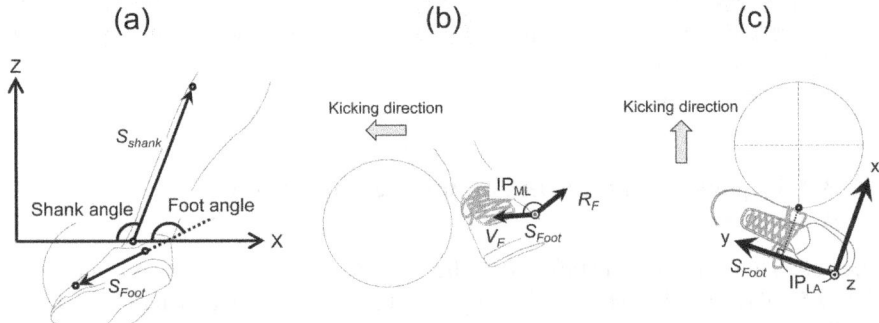

Figure 8.1 Definitions of the angles to express the ball impact characteristics. (a) Shank and foot inclination angle, (b) medial-lateral impact point, (c) impact point along longitudinal axis of the foot (modified from Shinkai & Nunome, 2008)

was also obtained. This was defined as the y component of the coordinates of the ball in the foot local coordinate system fixed at the lateral side of the calcaneus (Figure 8.1c). To clarify the ball impact characteristics of a knuckle shot, all impact parameters were compared to those of his maximal instep and side-foot kicks.

Of 10 consecutive attempted trials, a successful knuckle shot (Figure 8.2, trial A), which had remarkable change of ball trajectory and scored a goal, was obtained. Also another shot with similar initial conditions (Figure 8.2, trial B) was chosen for comparison. It seemed that the ball trajectory of trial A (with white marks) irregularly changed downward due to small ball rotations (0.8 rev/s).

The foot velocity just before ball impact, the initial ball velocity, and the shank inclination angle of knuckle shot were similar to those of his instep kicks (Table 8.1). These results showed he swung his kicking limb forward during the knuckle shot in a similar way as he performed a full instep kick. On the other hand, the foot inclination angle, IP_{ML}, and IP_{LA} were somewhere between his instep and side-foot kicking, as can be observed visually in Figure 8.3a. Specifically, in the case of knuckle shot, he hit the ball with the medial and proximal part of the foot (around the first metatarsal base), which was a hybrid position between that used for the instep and side-foot kicks (Figure 8.3b). Figure 8.4 shows the horizontal (XY) plane motion of the foot segment during ball impact. In the knuckle shot, the foot was more perpendicular to the kicking direction and moves forward more linearly than for instep kicking, though not to the extent of the side-foot kick. These results coincide with his remarks related to the ball impact of knuckle shots on a Japanese broadcasting program

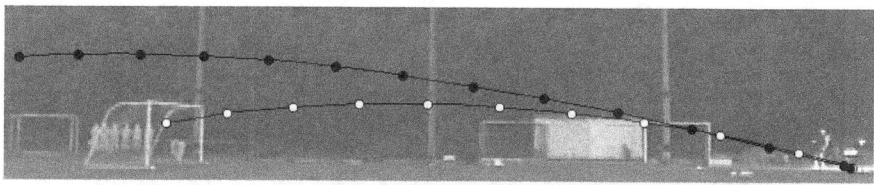

	Ball velocity (m/s)	Launch angle (deg)	Height passing through 25m ahead (m)		Number of ball rotations (rev/s)
			Theoretical value	Measured value	
Trial A	30.6	16.2	3.7	1.8	0.8
Trial B	30.6	15.3	3.3	4.4	1.4

Figure 8.2 Flying trajectories of knuckle ball kicking (○: trial A, •: trial B) and parameters of flying trajectory on knuckle ball kicks. Theoretical values of the height passing through 25 m ahead (goal line) were calculated using ball velocity and launch angle measured from high-speed video cameras (250 fps) considering air resistance (drag coefficients: 0.15, Asai et al., 2007) (modified from Shinkai & Nunome, 2008)

Table 8.1 Parameters related to ball impact of knuckle shot, instep kicking, and side-foot kicking.

	Foot velocity (m/s)	Ball velocity (m/s)	Inclination angle at impact (deg)		Foot contact point	
			Shank	Foot	IP_{ML} (deg)	IP_{LA} (cm)
Knuckle shot	20.9	30.6	122	174	151	9.7
Instep kicking	21.7	31.2	129	155	124	12.7
Sidefoot kicking	18.7	25.2	85	181	173	8.8

"hit the ball using between instep and infront of the part of foot", "push the centre of the ball forward linearly". However, only one remark "kicking with the image which contacts the ball as long as possible" does not agree with the fact: the contact duration of the knuckle shot is 9.0 ms, instep kicking is 8.5, and side-foot kicking is 9.0 ms. However, although Honda is a top professional footballer, he is yet not able to intentionally lengthen ball contact time (Nunome et al., 2014).

It can be assumed that as the ball reaction force during instep kicking was so large (more than 4 times of player's body weight) that the muscles around the ankle required for maintaining joint rigidity cannot exert the force required to control its joint motion (Shinkai et al., 2009). Nevertheless, the passive foot angular motions during the knuckle shot, especially plantar flexion, were reduced drastically compared to those of instep kicking (Figure 8.5). In the case of ball impact with the medial and proximal parts of the foot such as in the knuckle shot, the ball reaction force vector would act near the centre of gravity of the foot and

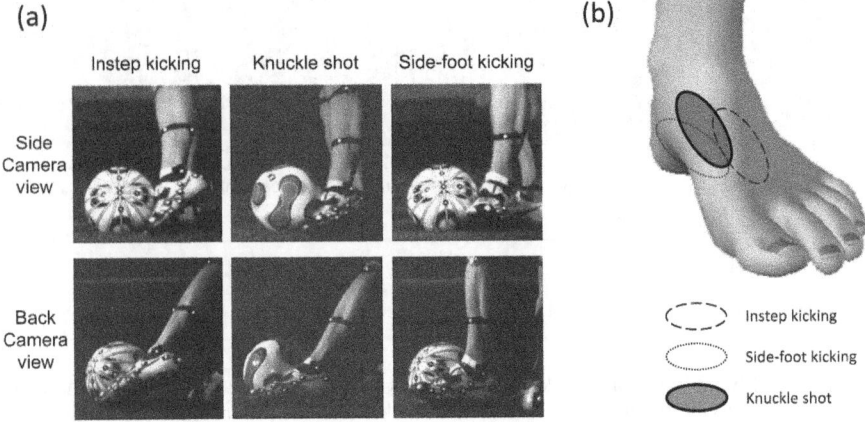

Figure 8.3 (a) Comparison of shank and foot posture at the instant of ball contact between three types of kick. (b) Comparison of foot impact point with the ball between three types of kick (modified from Shinkai & Nunome, 2008)

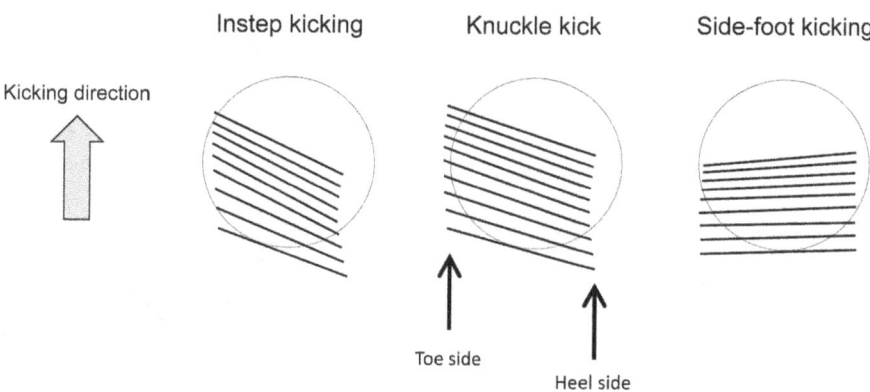

Figure 8.4 Motion of the foot segment to the ball during ball contact in the horizontal plane

would pass from the medial side to the lateral side of the foot. This action may allow the player to restrict passive foot motions to a minimum, especially plantar flexion, during ball impact.

Principally, to kick a non-spinning knuckle ball, the resultant force vector from the kicking foot must be maintained through the centre of gravity of the ball throughout ball impact. In the case of normal instep kicking, however, these passive and large foot motions most likely change the direction of this foot force vector (Figure 8.6a) and the 'gear effect' induces ball spin (mainly back spin) (Figure 8.6b). Therefore, minimising the passive foot motion during ball impact observed in a knuckle shot would seem to be a key to minimise the production of ball rotation. This would be the key coaching point that we would advise players to follow. From our results, it can be seen that a top professional player used a

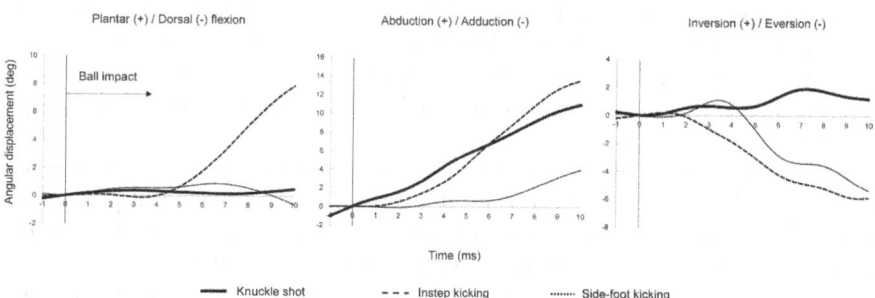

Figure 8.5 Three dimensional angular displacements of the ankle (plantar/dorsal flexion, abduction/adduction, and inversion/eversion) during ball contact between three types of kick. The angles of the foot at the instant of initial ball contact were defined as criterion values (zero degree)

(a)   (b)

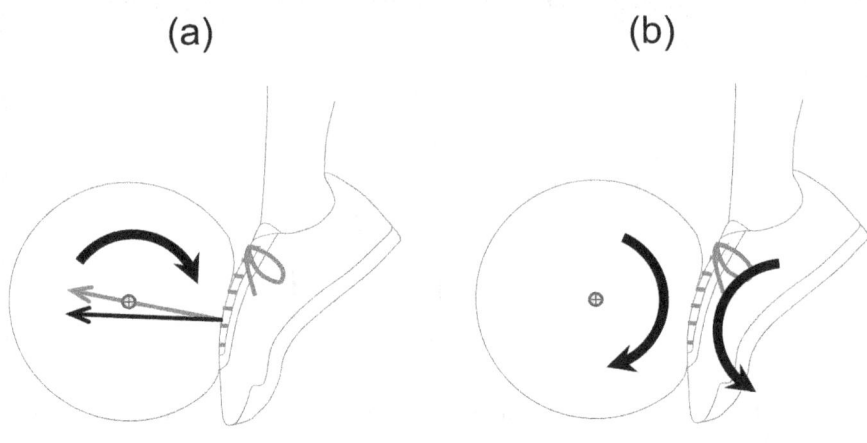

⊕ Centre of gravity of the ball

Figure 8.6 Diagrams of relationship between foot deformation and producing ball rotation during ball contact. The change of direction of foot force vector (a) and gear effect due to foot deformation (b)

very biomechanically effective ball impact technique to consistently launch a non- or low-spinning ball.

3. Comparison of ball impact technique between Keisuke Honda and Cristiano Ronaldo

As described in an earlier part of this chapter, a top Japanese professional soccer player Keisuke Honda used a biomechanically relevant, and an effective impact technique to produce a non- or low-spinning ball. It is also necessary to research the impact technique of a knuckle shot for other top level footballers in order to generalise Honda's technical characteristics. Therefore our views on how to impact the ball to enable a knuckle shot could be more generally applicable.

One of the world's most famous kicker of the knuckle shot is Cristiano Ronaldo. He is a player for Real Madrid in Spain, and at the time of press was also Portuguese national squad captain. Cristiano Ronaldo dos Santos Aveiro (age at the time 26yrs; height 1.85m; mass 80kg) was tested as part of a media investigation into his elite sporting capacity. Testing took place in an indoor television studio which was covered in 3G artificial turf. Ronaldo wore custom made Nike Mercurial Vapour SG. Ronaldo publically claims to have size 9 (US) feet, but it is widely rumoured that his soccer shoe size is smaller than this. A full size soccer goal was placed in the experimental volume, and a professional goalkeeper was instructed to attempt saves whilst remaining on the goal line. The stationary ball was placed 20m from the goal. Five curved free kicks were performed (side-foot/ instep), and five knuckle balls also performed (instep) with the aim of scoring a goal.

Ten Vicon T-Series cameras sampling at 250Hz were synchronised via a Vicon Giganet. A full body marker set was employed that had an upper body based on Vicon Plug-in-Gait, and a lower extremity marker set using the CAST system (Cappozzo *et al.*, 1996). This was chosen due to the restrictions on time, yet also maintaining the ability to gain accurate data on lower extremity functioning. A Coda pelvis marker set was used, along with medial and lateral markers on the knee and ankle, whose co-ordinates were held within marker triads placed in the centre of each segment. An extra foot marker was employed for tracking on the right (kicking) foot. Ball kinematics were measured using one Gigi-cam (Quintic Consultancy Ltd, UK) sampling at 250Hz in the sagittal plane. Kinematic measures of ball speed, trajectory, and ball spin were calculated from a sagittal plane view by Quintic ball spin software (Quintic Consultancy Ltd, UK). The best curved free kick (Ball velocity 25.5 m/s, Launch angle 20.5°, Total spin 7.67 rev/s, Top spin 6.80 rev/s, Draw spin 3.50 rev/s, Rifle spin anti-clockwise 0.62 rev/s), and knuckle shot (Ball velocity 29.4 m/s, Launch angle 13.4°, Total spin 1.35 rev/s, Top spin 0.10 rev/s, Fade spin 1.35 rev/s, Rifle spin clockwise 0.05 rev/s) were subjectively selected by the experimenter. Outputs from the software can be seen in Figure 8.7.

It is interesting to see the difference between the two styles of kick from Ronaldo. Firstly from the ball speed data we can see that the ball travels faster for the knuckle ball. This is required in order to exceed the Reynolds number

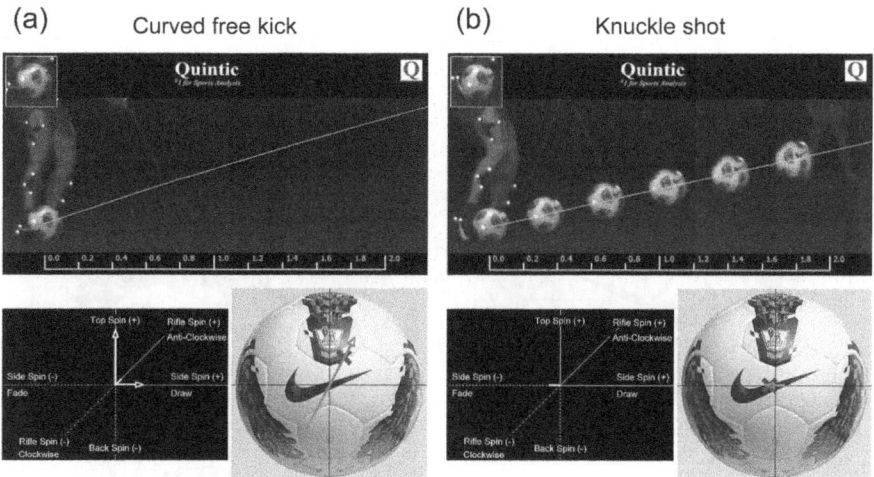

Figure 8.7 Ball launch, spin, and impact characteristics of Ronaldo curved free kick (a) and knuckle free kick (b)

required for the transition between turbulent and laminar flow in the ball boundary layer. It is to be noted that these kicks were performed in Television studio conditions, and as such the player did not generate ball speed as high as would be expected in a match situation. The quicker ball speed allows the flow characteristics at the boundary layer to change as the ball slows slightly during its flight path, and this, combined with the interaction of a very slowly rotating ball's seams providing a 'ramp' for the boundary layer that then causes the fluctuations we view during flight that is characteristic of a knuckle ball kick. During the curved free kick the flight path of the foot centre of mass progresses more medially to laterally, which creates the spin vector that we can see on the graphic. Naturally, increased spin comes at the expense of ball velocity. As shown in Figure 8.7b, the spin vector for the knuckle ball (which has very little spin) is almost through the centre of mass of the ball (estimated impact point on the 'x' symbol).

The support foot and kicking foot position in the horizontal (XY) plane were similar between curved free kick and knuckle ball kick. On the other hand, it is also interesting that there is far greater difference of the foot inclination angle in frontal (XZ) plane between two kicks. In fact a knuckle ball kick for this particular player has a replica plantar flexion angle that would be expected with a maximal instep kick, yet the impact point is more proximal (towards the ankle), meaning that the centre of mass of the foot is more closely aligned with the centre of mass of the ball (Figure 8.8). This appears to be the unique nature of Ronaldo's knuckle ball kick that allows him to generate high foot speed by the traditional sagittal plane hip flexion/knee extension, without having to externally rotate the kicking leg, and thus losing foot speed.

When comparing these two top player's data, it is necessary to keep in mind that two kicking experiments of Honda and Ronaldo were different in various

(a)   (b)

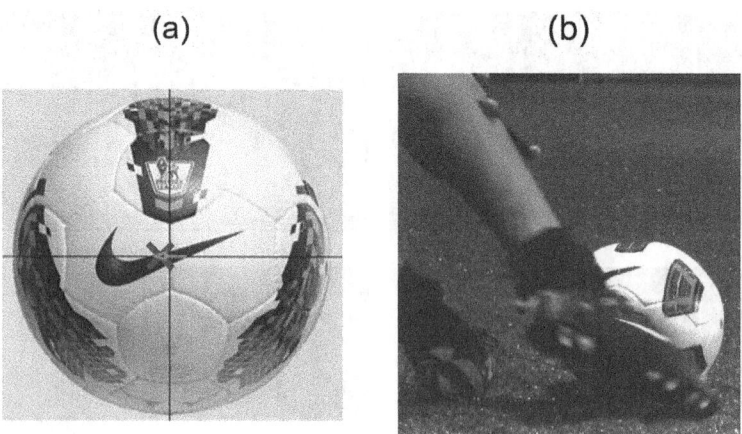

Figure 8.8 Ball impact of Ronaldo. (a) The hitting point of the ball. (b) The foot contact point with the ball at the instant of contact

aspects (especially differences in marker attachment position), and Ronaldo was not kicking with (perceived) maximal effort. Therefore a simple comparison between these two results is very difficult. However, as the data of the world's top player such as Ronaldo is very valuable, we will present these data and try to compare them with Honda as far as is possible.

Initial ball velocity, launch angle, and the number of ball rotations of Ronaldo's knuckle shot were 29.4 m/s, 13.4 degree, and 1.35 rev/s, respectively. In relation to the knuckle shot of Honda, the ball velocity was slightly lower and the number of ball rotation was higher (be equal to Trial B of Honda). The contact point of the ball was very close to centre of the ball (Figure 8.8). What is interesting about the ball impact of Ronaldo is the difference with that of Honda. Comparison of images at the instance of ball contact between Figure 8.3a and 8.8 showed that the foot inclination angle of Ronaldo in frontal plane was small (for the angle definition, refer to Figure 8.1a) and is comparable in magnitude to that of Honda's instep kick. These results indicate that the impact point of the knuckle shot of Ronaldo (IP_{ML}) was closer to that of his instep kicking than Honda's hybrid position. However, since it was observed that the impact point along longitudinal axis of the foot (IP_{LA}) was fairly close to the centre of gravity of the foot (coincident with Honda's), it can be suggested that bringing the foot impact point close to the centre of gravity of the foot (i.e. ankle) is essential factor for the success of knuckle shot. In addition, it is to be noted that the possible inclination angles available to players in the successful generation of a knuckle ball will be dependent on player anthropometrics, in this case mainly foot length. Those players with shorter foot length can use a smaller inclination angle whilst still maintaining the IP_{LA} close to the centre of gravity of the foot. This relatively small foot length exhibited by Ronaldo could well be why his foot position and impact points vary with the rest of the world's top players who also exhibit excellent knuckle ball kicks (Honda, David Luis, Drogba).

4. Application of their ball impact characteristics for knuckle shot training

The unique ball impact technique of top professional players mentioned above can provide useful information for amateur players who are training to acquire this new technique of producing the knuckle shot. To pilot test if Honda's ball impact technique can be applied in this way, we conducted a kicking training experiment.

A collegiate soccer player (age at the time 21 yrs; height 1.79 m; body mass 78.0 kg; shoe size US 8.5) participated in this case study. Prior to kicking training, the subject was provided with verbal and gestured instructions of the foot contact point of Honda, to 'hit the ball with the medial and proximal part of the foot, which was a hybrid position between instep and side-foot kicking', and the effectiveness of this for a knuckle shot. One session of knuckle shot training was composed of two parts. The first part was kicking the ball into the net about 20 m ahead to just embed the feeling of ball impact and ball spin (10 min). The Second part was kicking the ball to the goal 30m ahead over the wall (15 min). The subject was instructed to focus on the foot impact point and low ball spin throughout the training, and performed a total of 20 session (5 times / week). The ball used for the experiment was SPEEDCELL (Adidas Japan K.K., inflation: 1000 hPa). To verify the effects of the training, the second part of the training was replicated with more accurate biomechanical measures taken. Three ultra-high-speed video cameras (2000 fps) and a high-speed video camera (300 fps) were used for recording the kicking limb and ball motion around the ball impact phase, and calculation of ball spin rate respectively.

As shown in Figure 8.9, the foot impact point with the ball moved in closer to his ankle after training for 4 weeks. It was observed decreased IP_{LA} (from 12.7 ± 0.7 to 10.9 ± 0.5 cm) and increased IP_{ML} (from 147 ± 7 to 161 ± 3 deg.) after training: these results mean his hitting point shifted to more medial and proximal part of the foot (hybrid position between instep and side-foot kicking) and the foot position at ball impact changed as to hit the ball more squarely with the medial part as instructed. It can be proposed this change of hitting point lead to a great reduction of the number of ball rotations (from 2.5 ± 1.1 to 0.7 ± 0.2 rev/s), similar to that of Honda (0.8 rev/s). Moreover, the number of goals scored during the kicking experiment increased dramatically (from 0/14 to 7/20) as a result. Launch angle was also decreased (from 18.7 ± 2.6 to 15.3 ± 1.5 deg.), and this might have contributed to the increasing number of goals. Judging from a low value of ball rotation and its standard deviation, and high probability of scoring the goal, it might be said his knuckle shot reached a practical significant level for soccer game play. Foot velocity just before ball impact of post training was decreased compared to that of pre training (from 19.7 ± 0.9 to 18.3 ± 0.7 m/s). The reason for this decrease may probably have been caused by focusing more attention on the accuracy of ball impact after training. On the other hand, interestingly, ball velocity of post training was maintained the equivalent level as that of pre training (from 26.1 ± 1.3 to 25.8 ± 0.6 m/s): these results indicate the ball impact efficiency of his knuckle shot was improved throughout the 4 weeks kicking training. It is thought that decreasing the passive deformations of the foot and ankle during ball impact, caused by the change of the foot hitting point lead to this improvement of impact efficiency (Asami and Nolte, 1983). Ball velocities in both pre and post training were lower than that of Honda's knuckle shot (30.6 m/s). Since the player had to kick the ball over the wall 9.15m ahead, as in an actual soccer free kick and also in this experiment, kicking the ball into a goal is more difficult as ball velocity becomes greater. Moreover, since there is a trade-off between speed and accuracy in sporting actions, and it is very hard to be successful at both a fast swing of the kicking leg (i.e. fast ball velocity) and an accurate ball impact point. We therefore propose that this remarkable difference in the ball velocity is the most representative difference of knuckle shot performance between top professionals and amateur players.

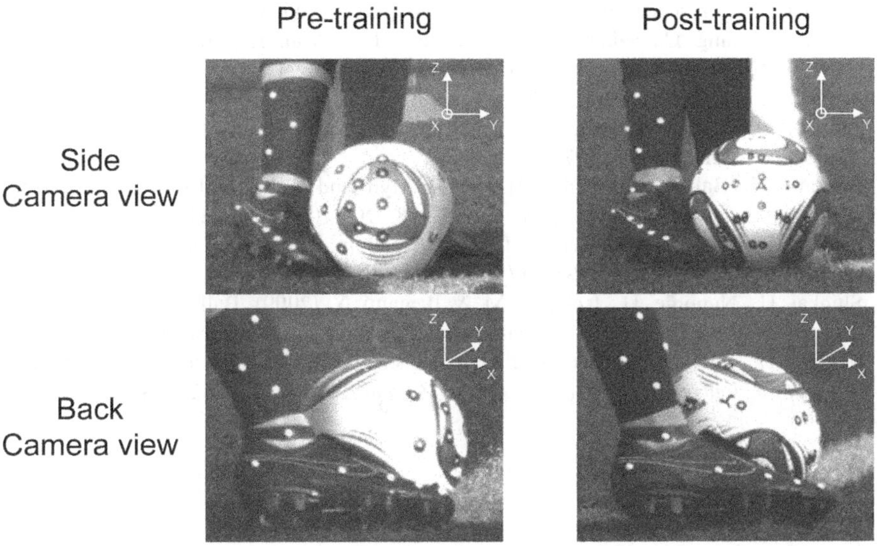

Pre-training   Post-training

Side
Camera view

Back
Camera view

Figure 8.9 Comparison of hitting point of the foot with the ball between pre and post specific knuckle shot training (modified from Shinkai & Nunome et al., 2008)

5. Conclusion

To execute with a non- or low-spinning kick, a top Japanese professional player hit the ball using a unique foot area which is a hybrid impact position between instep and side-foot kicks. Furthermore, using the comparison of the ball impact technique between this top Japanese player and the another world's top Portuguese player, it can be suggested that ball hitting technique close to the centre of gravity of the foot (in other words close to ankle) is an extremely important factor for the success of the knuckle shot. To establish and improve the performance of the knuckle shot, using these key points developed from biomechanical studies would be extremely useful as coaching cues in kicking training for the soccer player.

References

Asai, T. *et al.* (2007). Fundamental aerodynamics of the soccer ball. *Sports Engineering*, 10, 101–109.

Asai, T. & Kamemoto, K. (2011). Flow structure of knuckling effect in footballs. *Journal of Fluids and Structures*, 27, 727–733.

Asami, T. & Nolte, V. (1983). Analysis of powerful ball kicking. In H. Matsui, & K. Kobayashi (Eds.), *Biomechanics VIII-B*, Champaign (IL): Human kinetics, pp. 695–700.

Cappozzo, A., Catani, F., Leardini, A., Benedetti, M.G. & Della-Croce, U. (1996) Position and orientation in space of bones during movement: experimental artifacts. *Clinical Biomechanics*. 11 (2) 90–100.

Hong, S., Chung, C., Nakayama, M. & Asai, T. (2010). Unsteady aerodynamic force on a knuckle ball in soccer, In *The Engineering of Sport 8*, Wien: Elsevier, pp. 2455–2460.

Hong, S., Chung, C., Sakamoto, R., Nagahara, T. & Asai, T. (2013). A biomechanical analysis of the knuckling shot in football. In H. Nunome, B. Drust & B. Dawson (Eds.), *Science and Football VII*, London: Routledge, pp. 61–66.

Jinji, T., & Sakurai, S. (2006). Direction of spin axis and spin rate of the pitched baseball. *Sports Biomechanics*, 5, 197–214.

Nunome, H., Ball, K. & Shinkai, H. (2014). Myth and fact of ball impact dynamics in football codes. *Footwear Science*, 6, 105–118. DOI:10.1080/19424280.2014.886303.

Shinkai, H. (2010). The science of knuckle ball in soccer (in Japanese). *Journal of Training Science for Exercise and Sport*, 22, 299–306.

Shinkai, H., Nunome, H., Isokawa, M. & Ikegami, Y. (2009). Ball impact dynamics of instep soccer kicking. *Medicine and Science in Sports and Exercise*, 41, 889–897.

Shinkai, H., & Nunome, H. (2008). Ball impact of knuckle ball kicking (in Japanese). *Japanese Journal of Biomechanics in Sports and Exercise*, 12, 252–258.

Part III
Footwear

9 Kicking accuracy can be influenced by soccer shoe design

Ewald M. Hennig and Katharina Althoff

1. Introduction

The precision of passes and shots on goal is soccer players' most important kicking skill. Therefore, to win a game, kicking accuracy is much more important than being able to achieve maximum ball velocities. How much can the precision of kicks be influenced by external factors? Weather conditions, the type of playing ground, and the amount of air pressure in the ball are likely factors. We were interested in the question whether the type of footwear may also affect the precision of shots. Several studies were conducted to explore the possibilities of soccer footwear design for improving the precision of shooting balls.

Kicking accuracy is difficult to determine during soccer games. The players are under pressure to perform in competition situations such as one-against-one fights or kicks on goal. To study the kicking skills of players and/or the influence of external factors (footwear, ball pressure, ball spin, etc.), measurements will have to be performed under laboratory conditions. Several approaches were used to determine kicking precision by the number of successful shots on different kinds of targets (Katis *et al.*, 2013; Lees & Nolan, 2002; Teixeira, 1999). Other researchers (Chew-Bullock *et al.*, 2012; Figueiredo *et al.*, 2011; Kristensen, *et al.*, 2005; Nagasawa *et al.*, 2011) divided their targets into different sections for judging kicking precision. Rather than only detecting whether balls contacted a target or certain target areas, metric methods can be used to determine the actual distance from a target centre. Finnoff *et al.* (2002) used soccer ball imprints on a carbon paper covered wooden platform for measuring the distance between the target centre and the ball imprint locations. However, this metric method is laborious and time consuming, and does not allow the evaluation of many repetitive kicks. More recently, with the availability of digital videography and motion analysis systems, Scurr and Hall (2009), Young *et al.* (2010) and Alcock *et al.* (2012) filmed and analysed ball flight and its impact on a target at low data collection rates of 25 Hz. Sterzing *et al.* (2009) measured ball velocities and kicking accuracy for inside, instep, and full instep kicking techniques with 19 male soccer players. A target was constructed hanging a slightly transparent sheet down from the cross-bar of a 5 by 2 m goal construction. In the middle of the transparent sheet a bull's eye served as target centre. Using a high speed video camera (200 Hz) and a radar gun on the backside of the target, these researchers could determine ball velocities and impact locations on the target. As compared to the above mentioned videography studies the relative short distance of the camera towards the target and a high recording frequency guaranteed a better measuring accuracy. Although some video graphic studies used semi-automatic video data evaluation, it was still time consuming to load and analyse a video trace for each shot on the target. An economical ball accuracy measuring method with immediate results availability was introduced by Hennig *et al.* (2009). These researchers used conducting wires arranged in a circle on a wooden board. The impacting ball transferred electro-static charge to the circular wires, allowing the detection of its distance from the target centre. This method was used for the studies, described in this article. Therefore, this measuring method will be dealt with in more detail in the methodology section of this article.

Kicking accuracy studies, found in the literature, primarily focus on the skills of different soccer player groups (experienced and skilled vs. novice players, children and young vs. adult players, female vs. male players, etc.) and on the effects that various kicking techniques have on accuracy and speed. A possible influence of footwear, as it will be discussed in this article, was not analysed until recent years (Hennig *et al.*, 2009). The lack of interest in this topic may be caused by players' belief that kicking precision is skill based and cannot be influenced by their shoes. To find out which shoe properties players favour, we performed several surveys (Hennig, 2014). For each survey, players were asked about the most desirable features that they expect from their footwear. From a list of 11 shoe properties on a questionnaire the subjects ranked their five most important shoe characteristics from most (1) to least important (5). All shoe properties which were not chosen by the players received a score of '6'. The results of two surveys (1998, 2006) are summarized in Figure 9.1. In both surveys shoe comfort received by far the highest priority and it became even more significant to the players in 2006. Following comfort, these features were of similar importance to the players: ball touch, shoe stability, and traction. Kicking accuracy, kicking power, and ball spin production were given very low priorities. It is likely that our players believed that these features of ball handling are exclusively skill based. Soccer players probably do not believe that a shoe can improve their shooting accuracy and/or kicking speed. However, for maximum ball velocity, Sterzing and Hennig (2008) reported from a series of studies that maximum ball speed can be influenced by the type of shoe. Surprisingly, the players achieved highest ball velocities with their bare feet.

Although injury protection can be influenced by shoe design, its function of protecting against injuries also had a very low priority. Obviously, players are more interested in performance properties of their shoes rather than protection features. Modern low cut and light weight shoes follow this preference of players. They offer better performance on the field with little protection of the foot. An additional survey in 2006 with 69 female players demonstrated a very similar shoe priority ranking as seen for the men in Figure 9.1 (Althoff & Hennig, 2012).

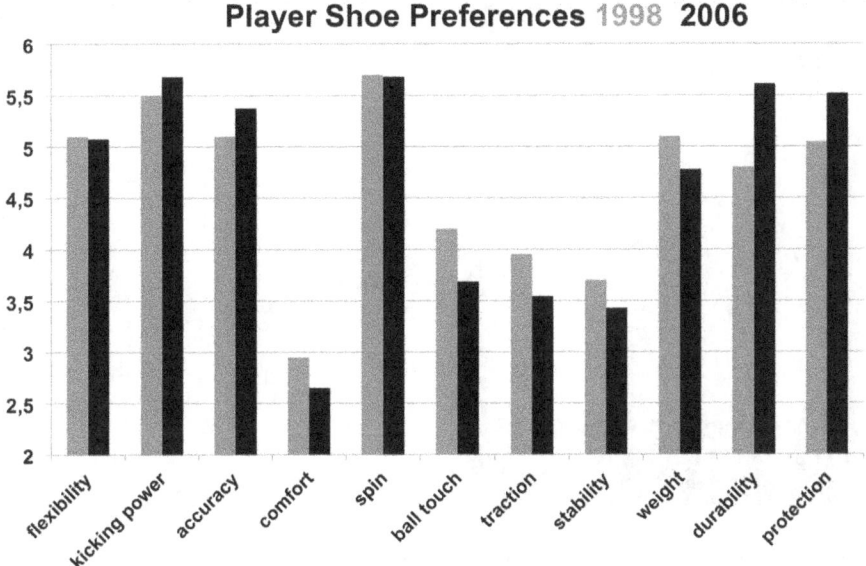

Figure 9.1 Priority ranking of desirable shoe properties (lowest value = highest priority)

2. Ball impact detection instrumentation

All studies, presented here, were performed with an electronic ball contact detection device. Ball acceleration during kicking causes friction between the ball, the ground, and the shoe. These friction phenomena generate electrostatical charging of the ball surface. In our laboratory the floor was covered with artificial turf, producing and transferring a substantial amount of charge to the ball. Because the ball material has insolating properties, but with moderate electrical resistance, the charges, generated by the friction with the ground and the shoe, are evenly distributed across the ball surface. Through its flight in the air, the electrical charge on the ball will remain and can be detected by contact by wires on a target. Following this idea, a circular electronic target with a diameter of 120 cm was built. Thirty electrically conducting wires were fastened in a concentric pattern on a wooden board. At contact with the board, electrostatic charges were transferred from the ball surface to the wires. The wires were connected to 30 electronic charge amplifiers. The voltage converted charge signals were then transferred from the electronic device at a frequency of 2 kHz per channel via A/D converter to a computer. Using a custom made program, the impact location was calculated as the weighted mean charge integral (during contact time) from the target strings. Immediately after the ball hit the target, the ball contact location was presented on the computer screen. The distance between the geometric centre of the target and the ball centre was chosen as accuracy measuring variable (in cm).

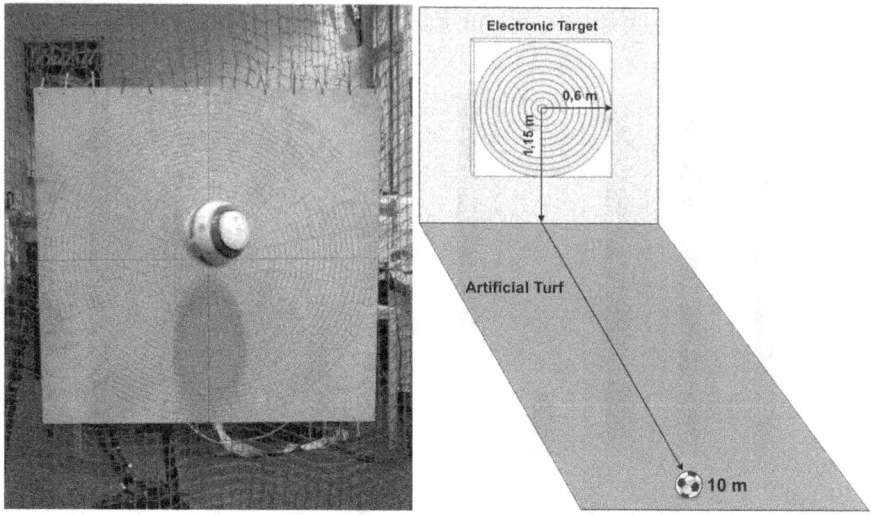

Figure 9.2 The electronic target arrangement in the laboratory

Measurements took place in the biomechanics laboratory of the University Duisburg-Essen with the target hanging from the ceiling. With this arrangement the centre of the electronic target was located at a height of 115 cm above the ground. The floor of the biomechanics laboratory was covered with artificial turf, and the experimental space was surrounded by soccer nets to protect the lab equipment from uncontrolled ball flights. Althoff *et al.* (2010) used this instrumentation to study the day-to-day reliability of accuracy measurements with 40 male subjects and three different shoe conditions. The authors concluded that for individual players the accuracy results show good repeatability from day to day. Furthermore, they found that kicking with unfamiliar conditions (assigned soccer shoe, barefoot) is as repeatable as kicking with their own, used shoes.

3. Results and discussion

Different kicking techniques are associated with more or less kicking accuracy (Lees & Nolan, 2002). Inside kicks are used for high precision short passes. However, inside kicks are not suited for high ball velocities. Instep kicks are less accurate but cover more distance and are primarily used for long passes and accurate shots on goal. Full instep kicks are least accurate but allow maximum ball speeds. Contact area size and surface geometry matching between foot and ball are mainly responsible for the precision of shooting. Independent from these skill based influences on kicking accuracy we wanted to explore whether the type of footwear can also have an influence on shooting precision.

3.1 Influence of shoe construction on kicking precision

To get an idea whether shoes may have an effect on the accuracy of kicks, we performed a study with five commercially available soccer shoes and also barefoot kicking (Hennig *et al.*, 2009). From a distance of 10 m, 24 male subjects were instructed to perform best accuracy kicks to the centre of the electronic target at a height of 115 cm above the ground. To guarantee best accuracy shots from each player no restrictions were imposed on either kicking technique or ball speed. However, the subjects were instructed to maintain their chosen kicking strategy in all six measurement conditions. For each experimental condition all players performed 20 repetitive kicks. The assignment of shoe conditions was randomised. The commercially available shoe models in this study were: Adidas Predator, Adidas F50 Tune-it, Nike Air Legend FG, Nike Total 90 FG Synthetic, Nike Total 90 FG Leather. All shoes were firm ground (FG) models to guarantee good traction on the artificial turf covering the laboratory floor. For the barefoot condition, socks were worn on the kicking foot to avoid skin pain as a consequence of friction between ball and skin. Wearing the socks our subjects felt comfortable in shooting the ball on the target.

The mean accuracy of the 20 repetitive kicks was determined for the 24 subjects in each of the six experimental conditions. Figure 9.3 shows the averaged distance of ball impacts from the target centre for the barefoot shots and kicking in the five soccer shoe models (ANOVA, $p < 0.01$). For the barefoot kicks a statistically significant reduced accuracy was found against all five shoe conditions. This was unexpected because the touch on ball was certainly best during the barefoot shots. The foot was only covered by a sock, thus providing the best proprioceptive feedback from the mechanoreceptors of the skin. In one of the

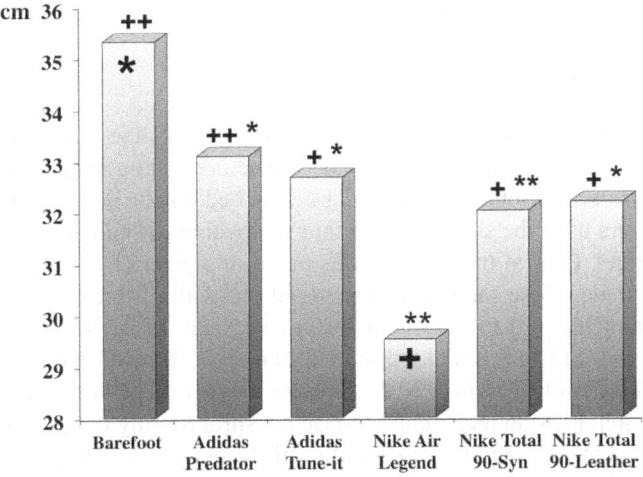

Figure 9.3 Kicking accuracy – mean distance from target centre (+* $p < 0.05$, ++**
$p < 0.01$)

footwear conditions (Nike Air Legend) the players showed a better kicking accuracy ($p<0.01$, $p<0.05$) as compared to the other four shoes. The models that covered the lacing area (Adidas F50 Tune-it) or used laterally oriented oblique lacing (both Nike Total 90 models) performed worse against the Nike Air Legend with its traditional straight lacing. Covering the lacing area and/or using oblique lacing was introduced by the manufacturers to provide a smooth contact between the ball and shoe contact surface for instep and full instep kicks. Thus, it could have been expected that these shoe models would perform better. From the differences in design between the shoes and considering the precision advantage of the Nike Air Legend, we were not able to identify shoe features that could have caused the observed difference in kicking accuracy between shoes. To follow up on the above results we suggested a few hypotheses that could be responsible for the effect of an enhanced precision of kicking with shoes:

- Better touch of the ball may enhance accuracy. However, we found that barefoot kicks had the least precision. Therefore, we did not follow up on ball touch.
- An increase in shoe mass on the kicking leg causes a higher moment of inertia, thus stabilizing foot swing. This could result in a better shooting accuracy. However, lighter shoes allow faster movements on the field. Therefore, adding additional weight to the shoe for better shooting precision will be counterproductive in modern soccer games.
- More friction between the shoe upper and the ball will prevent slipping of the ball during contact and thus cause better precision.
- A more homogenous pressure distribution between the ball and the shoe upper will enhance kicking precision.

The last two items of this list were topics of our further studies.

3.2 Shoe friction properties

To study the effect of friction, the upper of a Nike Total 90 FG shoe model was modified. The upper material friction and surface structure of the shoes were modified with 4 different shrouds that could be attached to the shoe upper. Eighteen soccer players used the shoe with the four removable shrouds for accuracy shots on the target. One of the shrouds was made of leather (L), the other three shrouds were made of the same rubber material but with different surface structure patterns from fine (F) to intermediate (I) and rough (R). The players performed 20 repetitive full instep kicks from a distance of 10 m on the target with its centre at a height of 115 cm. For this accuracy experiment the subjects had to use the full instep kicking technique because the shrouds only covered the surface of the shoe making contact with the ball for this technique. The three high friction rubber covers demonstrated a lower kicking accuracy against the leather cover result. Therefore, an increase in friction to reduce slipping between the ball and the shoe upper apparently has no influence on the precision of shots. It may even be detrimental because the leather surface showed the best results. For the three rubber cover conditions there is a trend towards better kicking precision with smoother surfaces from "Rough RS" to "Fine RS". The smooth leather surfaces and "Fine RS" with a delicate surface structure showed the best results. It is possible that uneven structures have a detrimental effect on kicking accuracy by causing uneven pressures across the ball contact area with the shoe. This may also explain the fact that barefoot precision shots exhibit the worst accuracy (Figure 9.3). The anatomical foot structures (bony prominences of the foot) may cause high pressure gradients and uneven pressures across the area of ball contact. To test this hypothesis a study was conducted to analyse the pressure distribution between shoe and ball during kicking.

3.3 Pressure distribution between shoe and ball

Based on the above mentioned potential influences for better kicking accuracies the factors of friction and ball touch seem to have no or only a minor influence. The friction experiments revealed however, that a dishomogenous pressure distribution between shoe upper and the ball caused by uneven surface structures may worsen shooting precision. The effect of a more homogenous pressure distribution on the accuracy of ball flight is also observed in tennis. Tennis players know that the tension strength of strings across the racket head influences accuracy as well as tennis ball speed (Hennig, 2007). During ball impacts, string deformation causes a lower energy loss as compared to the energy lost in ball deformation (trampoline effect). Therefore, lower string tensions will lead to higher ball velocities. On the contrary, an increase in racket string tension results in lower ball speeds but improved ball control. This is caused by a larger contact area and a more uniform pressure distribution between the strings and the ball during the impact. A similar principle may also apply to the interaction of shoes with the ball in soccer. Therefore, we conducted a study to measure the pressure distribution between ball and shoe during kicks for best accuracy. For measuring the dorsal pressures between the shoe upper and the ball a Pedar (Novel Inc., Munich) pressure distribution measuring insole was used on top of the shoes. With an elastic rubber band the insole was positioned and fastened to the medial mid- to forefoot area of the shoe, where players hit the ball during instep kicks. According to Nunome *et al.* (2006) the contact duration between foot and ball during instep kicks is very short (9 to 12 ms). To achieve a high sampling rate of almost 600 Hz only 35 out of 99 available insole pressure sensors were selected and used for the measurements. Twenty male soccer players performed instep accuracy kicks on our electronic target with its centre at a height of 115 cm. Twenty repetitive kicks were carried out in two footwear conditions from a distance of 10 m onto the target. Two shoe models from our first study (Nike Air Legend FG, Nike Total 90 FG Synthetic) that showed a statistically significant difference in kicking accuracy, were chosen for this experiment. The attachment of the two pressure measuring mats on the shoe upper of both shoes had to be performed carefully. By online monitoring of the pressure signal, when pushing against various anatomical foot landmarks, it was made sure that identical pressure sensors on the two pressure mats matched the same anatomical locations of the foot in the two shoe models.

Although the shoe uppers of both shoes were covered with the Pedar pressure measuring pad the accuracy results confirmed our previous finding of a better accuracy ($p < 0.05$) for the Nike Air Legend (Figure 9.3). From the pressure distribution measurements the peak pressures for each sensor and all repetitive kicks were evaluated. Figure 9.4 shows the averages of the peak pressures from all players and all of their target kicks in the two footwear conditions. The centre of the peak pressures across the 35 transducers is indicated by dark circles in the graphs. For the better accuracy Nike Air Legend the centre of pressure is located more medially and posterior on the foot dorsum.

Summing the peak pressure differences between all adjacent transducer, the Nike Total 90 value exceeded the Nike Air Legend sum by almost 20%. Higher differences in pressure values from transducers to their adjacent neighbours imply less homogeneity of the load distribution. Therefore, our results confirm that a more homogenous pressure distribution between the shoe upper and the ball is the key factor for a better kicking accuracy. As mentioned before, this result coincides with observations from tennis. More homogenous pressures between a tennis ball and the racket head strings guarantees higher accuracy and better ball control.

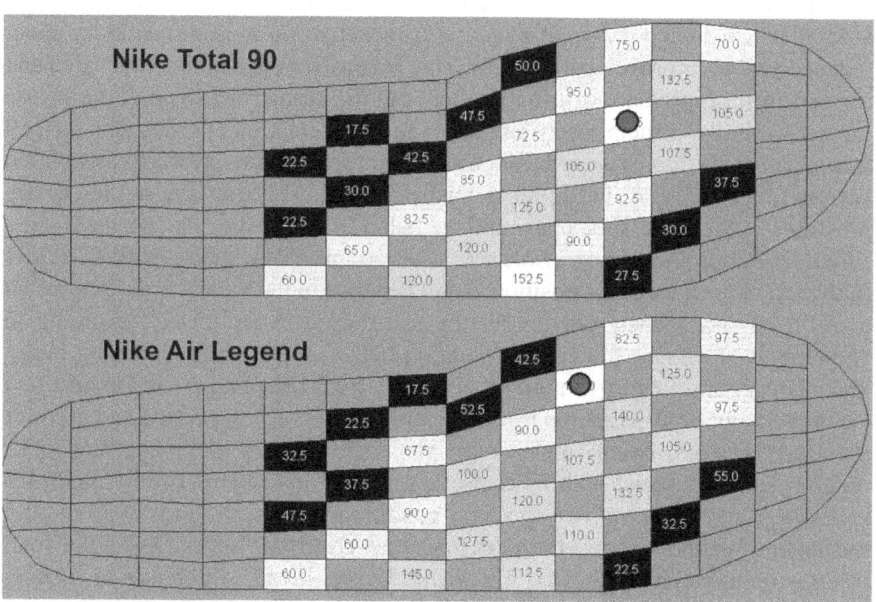

Figure 9.4 Averaged peak pressures (N/cm²) between shoe and ball for the shoes with less (Nike Total 90) and better (Nike Air Legend) kicking precision

3.4 Modifications for improving pressure homogeneity

Originating from the finding that a more homogenous pressure distribution can increase kicking accuracy we were motivated to modify shoes for better precision. However, footwear modifications are not easy to do and are costly. Therefore, we decided to perform padding of the foot in order to reduce the effect from the bony prominences. Padding (Polyurethane, 6 mm, 10 Shore A) was attached along the medial foot for smoothening the transition of the 1st metatarsal bone along the longitudinal arch of the foot. Furthermore, the gap between toe knuckles was filled with another padding (Dr. Scholl Silicone Gel Pad). Eighteen male soccer players performed 15 repetitive accuracy shots on the target from a distance of 8 m. Using a Nike Vapor shoe model the subjects performed with their right foot with and without padding instep and full instep precision kicks on the electronic target. As expected from the used kicking technique the full instep kicks showed lower precision. Although statistically not significant, there was a trend towards better accuracy for the instep shots with the foot padding (p = 0.11). Similarly, there was also a trend for better precision with padding in the full instep kicks (p = 0.12).

4. Conclusions

Most soccer players probably do not believe that their shoes can help them to perform more accurate short and long passes as well as shots on goal. In a series of studies we could demonstrate that kicking accuracy can be influenced by footwear. Considering various influences that may contribute to shooting precision the homogeneity of the pressure distribution between shoe upper and ball was identified as key factor. As compared to shod kicking the bony prominences of the unshod foot create inhomogeneous pressures across the area of ball contact, resulting in low kicking accuracy. It was also shown that shoes can be modified for better shooting precision. Pressure distribution instrumentation as well as the use of an electronic target are valuable tools to improve the accuracy characteristics of soccer shoes during the design process.

References

Alcock, A. M., Gilleard, W., Brown, N. A. T., Baker, J., & Hunter, A. (2012). Initial ball flight characteristics of curve and instep kicks in elite women's football. *Journal of Applied Biomechanics, 28*(1), 70–77.

Althoff, K., Hennig, E. M., Batz, S., & Bürgel, D. (2010). Day to day reliability of kicking accuracy in soccer. In: *Proceedings of the 16th Bianual Conference of the Canadian Society for Biomechanics* (pp. 51), Queen's University, Kingston, Ontario, Canada.

Althoff, K., & Hennig, E. M. (2012). What female and male soccer players expect from their shoes. In: *Proceedings of 3rd World Conference on Science and Soccer* (pp. 82). Ghent, Belgium: Victoris & Gent BC.

Chew-Bullock, T. S., Anderson, D. I., Hamel, K. A., Gorelick, M. L., Wallace, S. A., & Sidaway, B. (2012). Kicking performance in relation to balance ability over the support leg. *Human Movement Science, 31*(6), 1615–1623. doi:10.1016/j.humov.2012.07.001

Figueiredo, A. J., Coelho e Silva, M. J., & Malina, R. M. (2011). Predictors of functional capacity and skill in youth soccer players. *Scandinavian Journal of Medicine and Science in Sports, 21*(3), 446–454. doi:10.1111/j.1600-0838.2009.01056.x

Finnoff, J. T., Newcomer, K., & Laskowski, E. R. (2002). A valid and reliable method for measuring the kicking accuracy of soccer players. *Journal of Science and Medicine in Sport, 5*(4), 348–353.

Hennig, E. M. (2007). Influence of racket properties on injuries and performance in tennis. *Exercise and Sport Science Review, 35*(2), 62–66. doi:10.1249/JES.0b013e31803ec43e

Hennig, E. M. (2014). Plantar pressure measurements for the evaluation of shoe comfort, overuse injuries and performance in soccer. *Footwear Science, 6*(2), 119–127. doi:10.1 080/19424280.2013.873486

Hennig, E. M., Althoff, K., & Hömme, A. K. (2009). Soccer footwear and ball kicking accuracy. *Footwear Science, 1*(Supplement 1), 85–87.

Katis, A., Giannadakis, E., Kannas, T., Amiridis, I., Kellis, E., & Lees, A. (2013). Mechanisms that influence accuracy of the soccer kick. *Journal of electromyography and kinesiology: Official journal of the International Society of Electrophysiological Kinesiology, 23*(1), 125–131. doi:10.1016/j.jelekin.2012.08.020

Kristensen, L. B., Andersen, T., & Sorensen, H. (2005). Comparison of precision in the toe and instep kick in soccer at high kicking velocities. In: T. Reilly, J. Cabri, and D. Araújo (Eds.), *Science and Football V* (pp. 70–72). London: Routledge.

Lees, A., & Nolan, L. (2002). Three-dimensional kinematic analysis of the instep kick under speed and accuracy conditions. In: W. L. Spinks, T. Reilly, and A. Murphy (Eds.), *Science and football IV* (pp. 16–21). London: Routledge.

Nagasawa, Y., Demura, S., Matsuda, S., Uchida, Y., & Demura, T. (2011). Effect of differences in kicking legs, kick directions, and kick skill on kicking accuracy in soccer players. *Journal of Quantitative Analysis in Sports, 7*(4). doi:10.2202/1559-0410.1339

Nunome, H., Lake, M., Georgakis, A., & Stergioulas, L.K. (2006). Impact phase kinematics of instep kicking in soccer. *Journal of Sports Sciences, 24*(1), 11–22.

Scurr, J., & Hall, B. (2009). The effects of approach angle on penalty kicking accuracy and kick kinematics with recreational soccer players. *Journal of Sports Science and Medicine, 8*(2), 230–234.

Sterzing, T., & Hennig, E. M. (2008). The influence of soccer shoes on kicking velocity in full-instep kicks. *Exercise and Sport Science Review, 36*(2), 91–97. doi:10.1097/ JES.0b013e318168ece7

Sterzing, T., Lange, J., Wächtler, T., Müller, C., & Milani, T. (2009). Velocity and accuracy as performance criteria for three different soccer kicking techniques. In: A. J. Harrison, R. Anderson & I. Kenny (Eds.) *Proceedings of the 27th International Society of Biomechanics in Sports* (pp. 243–246), University of Limerick, Limerick, Ireland.

Teixeira, L. A. (1999). Kinematics of kicking as a function of different sources of constraint on accuracy. *Perceptual and Motor Skills, 88*, 785–789.

Young, W., Gulli, R., Rath, D., Russell, A., O'Brien, B., & Harvey, J. (2010). Acute effect of exercise on kicking accuracy in elite Australian football players. *Journal of Science and Medicine in Sport, 13*(1), 85–89. doi:10.1016/j.jsams.2008.07.002

10 Biomechanical and motor performance differences between female and male soccer players

Influences on the game and recommendations for gender specific footwear

Katharina Althoff and Ewald M. Hennig

1. Introduction

Women's soccer has become increasingly popular in recent years. The international soccer association FIFA reported a 37% increase in female soccer players from 21.9 million in the year 2000 to 30 million in 2014 (FIFA, 2007, 2015). Nevertheless, previous research mostly addressed men's soccer with little consideration of gender issues.

There are many anatomical and physiological, psychological, and constitutional differences between adult women and men: on average women are 10 to 15 cm shorter and 10 to 20 kg lighter than men. In comparison to men, women have a wider pelvis and smaller shoulders which leads to a lower centre of mass (O'Brien, 1995; Wilmore & Costill, 1994). The wider pelvis also often causes a valgus alignment of the knees that could increase knee injury risks (Gokeler *et al.*, 2010; Horton & Hall, 1989). Women have a lower percentage of fat-free mass related to body weight (62 to 66%; men: 73 to 76%) (Frontera *et al.*, 1991; Lindle *et al.*, 1997; Neder *et al.*, 1999) and a lower percentage of muscle mass related to body weight (25 to 23%; men: 40 to 45%) (Hollmann & Hettinger, 2000; Thein & Thein, 1996). Due to the lower muscle mass women have a lower absolute muscle strength: for the lower extremities (hip flexors/extensors and lower leg flexors) women achieve about 80% of men's strength and across all muscle groups they achieve on average 66 to 70% of men's strength (Drinkwater, 1989; Hollmann & Hettinger, 2000). Some cardio-vascular characteristics point towards a lower physiological efficiency in females. The average absolute heart volume of untrained women is about 500 to 600 ml (men: 600 to 800 ml) and the relative heart volume is about 9.5 to 10 ml/kg body weight (men: 11 to 12 ml/kg). The lower female heart volume leads to an uneconomical increase of heart rate during physical stress (König *et al.*, 1961; Roskamm *et al.*, 1961; Weineck, 2010). Furthermore, women have less blood volume (women: 3.8 l; men: 5.1 l), lower haematocrit values (women: 37 to 47%; men: 40 to 54%) and lower haemoglobin contents (women: 12 to 16 g/dl; men: 13 to 18 g/dl) (Åstrand *et al.*, 1973;

O'Brien, 1995; Thein & Thein, 1996; Weineck, 2010). The absolute as well as relative respiratory minute volume is also lower in women as compared to men (absolute: 90 ml/min compared to 110 ml/min; relative: 1.5 l/kg compared to 1.6 l/kg) (Åstrand et al., 1973; Schmidt et al., 2010; Thein & Thein, 1996; Weineck, 2010). Resulting from the above physiological variables, women have a lower maximum oxygen uptake und therefore a lower aerobic capacity.

These biological differences directly affect match performance of female and male soccer players in general and soccer specific movements and ball handling skills. Detailed knowledge about the different playing and moving patterns is important to develop gender specific training concepts and sports equipment. Footwear should be made available to women that matches the specific needs of female soccer players as best as possible.

2. Playing behaviour

Biological differences lead to different playing behaviours of female and male soccer players. Moreover, the level of professionalization is lower in women soccer. This influences also the women's playing behaviour as a result of different training modalities, medical treatments or match preparations. In the following sections on the evaluation of high performance women's and men's soccer matches, differences in running performance, match activities and used kicking techniques are described.

2.1 Running performance

The total distance of female soccer players covered during a match is about 9.9 to 10.8 km. Even though women show considerably lower aerobic endurance capacities, they cover only slightly lower distance during a soccer match compared to male soccer (10.4 to 11.1 km). However, the high intensity run results show clear gender differences: Women perform a lower number of sprints (women: 26; men: 39) and cover substantially lower distances with high running velocities (women: 1.31 km; men: 2.43 km) and sprinting (women: 0.16 km; men: 0.65 km) (Andersson et al., 2008; Bradley et al., 2014; Krustrup et al., 2005; Mohr et al., 2003).

2.2 Match activities

Althoff et al. (2010) identified gender specific match activities by using video analyses of the women's World Cup 2003 and the men's World Cup 2002. Overall the number of passes, ball controls and shots was just 8% lower in the women's compared to men's game. Based on net playing time, female players performed 11.1 and male players 14.4 actions with the ball per minute. To identify game patterns, data were normalized to the total number of actions. Compared to men, women performed long passes more frequently (above 25 m) whereas men used short range activities more often, such as ball controls and short passes (see Figure 10.1). In 2013, data from the women's World Cup from 2011 were compared to the data from the 2003 World Cup to analyse changes in the women's game (Althoff *et al.*, 2013). In general, there was a trend for the women's game to become similar to the men's game: The number of long passes decreased and the amount of short range activities increased in 2011 (see Figure 10.1).

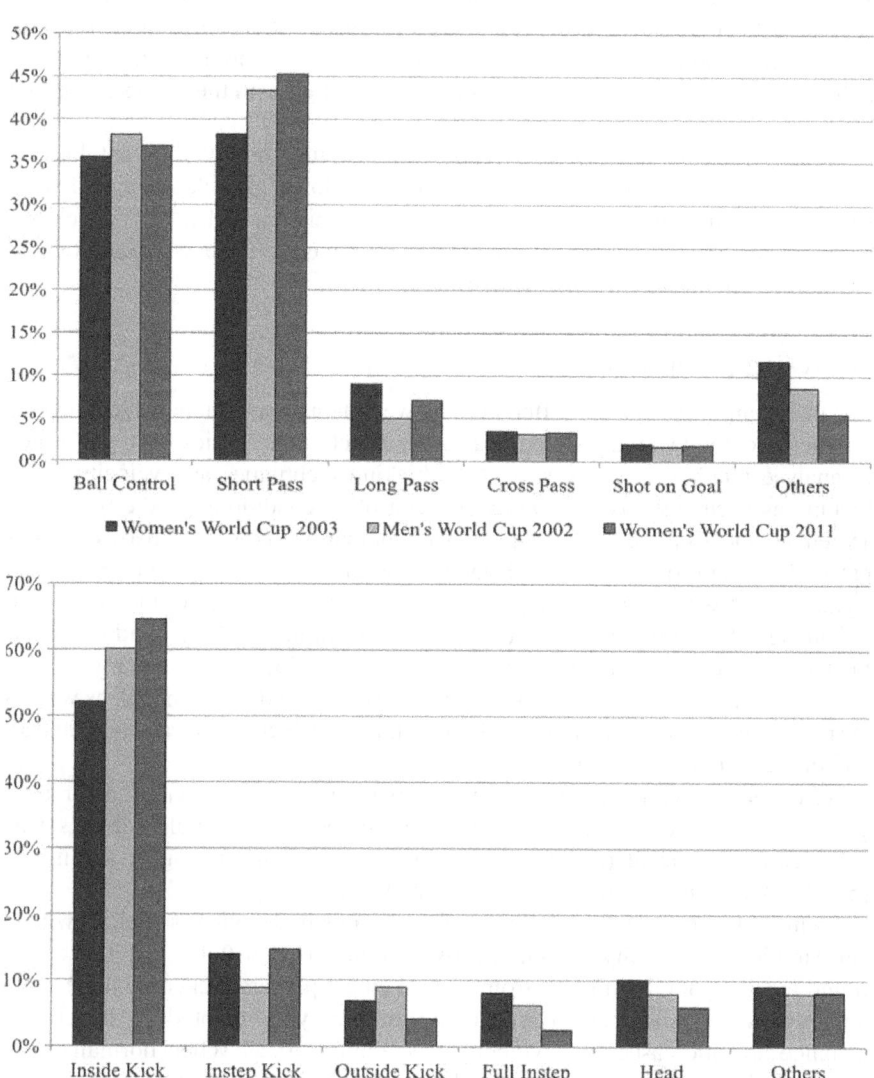

Figure 10.1 Percentile distributions of ball actions and kicking techniques

2.3 Kicking techniques

An analysis of the kicking techniques for all ball actions showed a more frequent use of the instep kick in both women's World Cups compared to the men's World Cup. Furthermore, the female players in 2011 used the inside much more often for kicking, passing and controlling the ball than the female players in 2003 (see Figure 10.1). Even for the same actions with the ball, gender differences as well as developments in the women's game were observed: Female players in 2003 – compared to male players in 2002 – used the instep kick more often for long passes (73% compared to 66%) and women in 2011 – compared to women in 2003 – passed the ball over a short distance more often with the inside of the foot (76% vs. 64%).

Kicking velocity and kicking accuracy are influenced by different kicking techniques (Sterzing *et al.*, 2009a). Due to their lower muscle strength female soccer players have to use kicking techniques that achieve a higher kicking velocity to cover the same distances as male soccer players. However, these kicking techniques demonstrate a lower accuracy.

3. Kicking performance

The match analysis showed differences between female and male players regarding their kicking techniques used in a soccer match. Two studies were performed to analyse kicking accuracy as well as kicking techniques and velocity while kicking as accurately as possible under controlled conditions. In the first study 18 female and 18 male soccer players with comparable playing expertise participated. They had to kick from a distance of 8 and 11 m on an electronic target, trying to hit the centre of the target with best accuracy. The electronic target with a diameter of 1.20 m and a centre height of 1.15 m above the ground measured the kicking accuracy as distance from the ball impact location to the target centre (Hennig *et al.*, 2009). Furthermore, velocity was detected for each kick as well as the maximum kicking velocity for each subject. Using a high-speed camera (600 fps), kicking techniques were characterized.

As expected, with increasing distance kicking accuracy decreased for both genders. The difference in accuracy between female and male subjects was statistically significant ($p < 0.05$) at a distance of 8 m and highly significant ($p < 0.01$) at a distance of 11 m (see Figure 10.2).

From a distance of 8 m both the female and male players kicked with a similar kicking velocity. A statistically significant ($p < 0.05$) difference was found for the velocity of kicks from 11 m: Female players kicked with the same velocity as they did from 8 m, but the male players adapted to the greater distance and increased their velocity (see Figure 10.2). When normalized to their individual maximum kicking velocity, female players kicked with a much higher relative velocity. From a distance of 8 m they kicked with 77% of their maximum kicking velocity whereas the male players only needed 59% of their maximum kicking velocity.

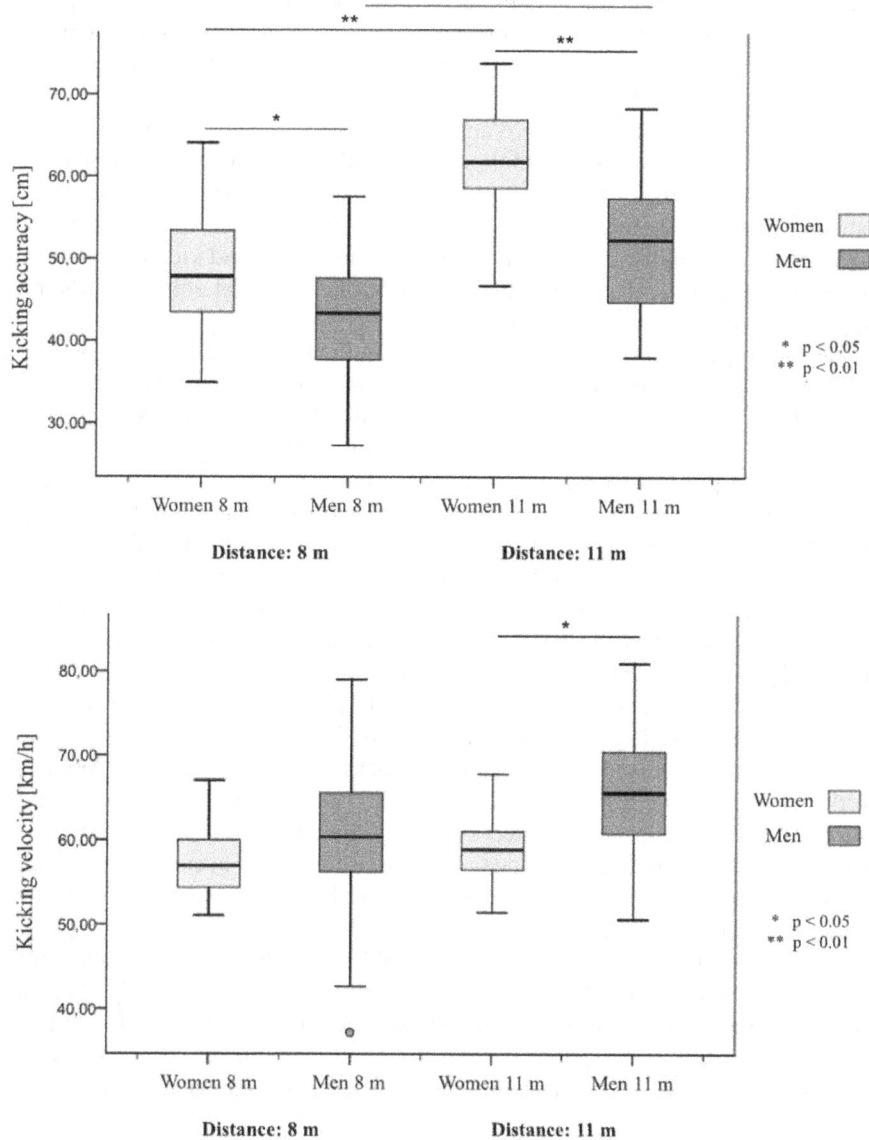

Figure 10.2 Kicking accuracy and velocity (for kicking as accurately as possible) from two distances

No gender differences were found for the used kicking techniques at the two distances; both genders used predominantly inside kicks. However, female players kicked less frequently with a stable ankle joint fixation. These kicks with an unstable ankle tended to be less accurate ($p = 0.065$) and were statistically significantly slower ($p < 0.05$).

A second study was performed to analyse the use of kicking techniques from greater distances. Twenty-one female and 25 male recreational soccer players with similar experience and performance took part in this study. They kicked from six different distances (8 to 23 m) at a soccer goal (2.44×7.32 m) that was divided into 15 zones. Their aim was to hit the central zone and kicking accuracy was determined as the distance from the middle of the central zone towards the centre of the zone that was hit by the ball.

Distance to goal had a statistically highly significant ($p < 0.01$) influence on kicking accuracy in both subject groups, whereas female soccer players kicked less accurately with increasing distance. Up to the distance of 14 m no gender differences were found regarding the used kicking techniques: Female and male players kicked predominantly with the inside foot. At a distance of 17 m the female subjects kicked most often with their instep (62%) and used full instep kicks (21%) more frequently than inside kicks (17%). From the same distance the male soccer players used the full instep kicks rarely (3%) but had high percentages of inside (46%) and instep kicks (51%) as shown in Figure 10.3. Over all distances, the men used the inside kicking technique more often, whereas the female players had higher percentages of instep and full instep kicks.

Compared to the men, female soccer players have to use different kicking techniques to achieve high distances on the field. Because of their lower muscle strength they have to use kicking techniques that result in higher ball velocities, but these techniques are less accurate. A gender specific soccer training should take the more frequent use of instep kicks in female players into account. This technique should be trained in a special manner to improve accuracy. However, the women should also focus on achieving higher ball velocities with the inside technique, because this technique guarantees the best accuracy.

Furthermore, female soccer players should strengthen their muscles for stabilizing the ankle joint during ball contact. The first accuracy study showed that female players kick less often with a stable ankle and that these kicks are less accurate. Therefore, a specific strength training, combined with a kicking technique training, should result in improved kicking accuracy and velocity.

4. Soccer shoes influence on running performance

The different use of kicking techniques characterizes gender specific playing behaviour. Furthermore, playing performance is influenced by the speed of movement. Due to the above mentioned biological differences, women are less powerful and slower than men. The soccer shoe plays an important role for achieving high movement speeds. During a soccer game many rapid acceleration and deceleration motions as well as changes of direction have to be performed.

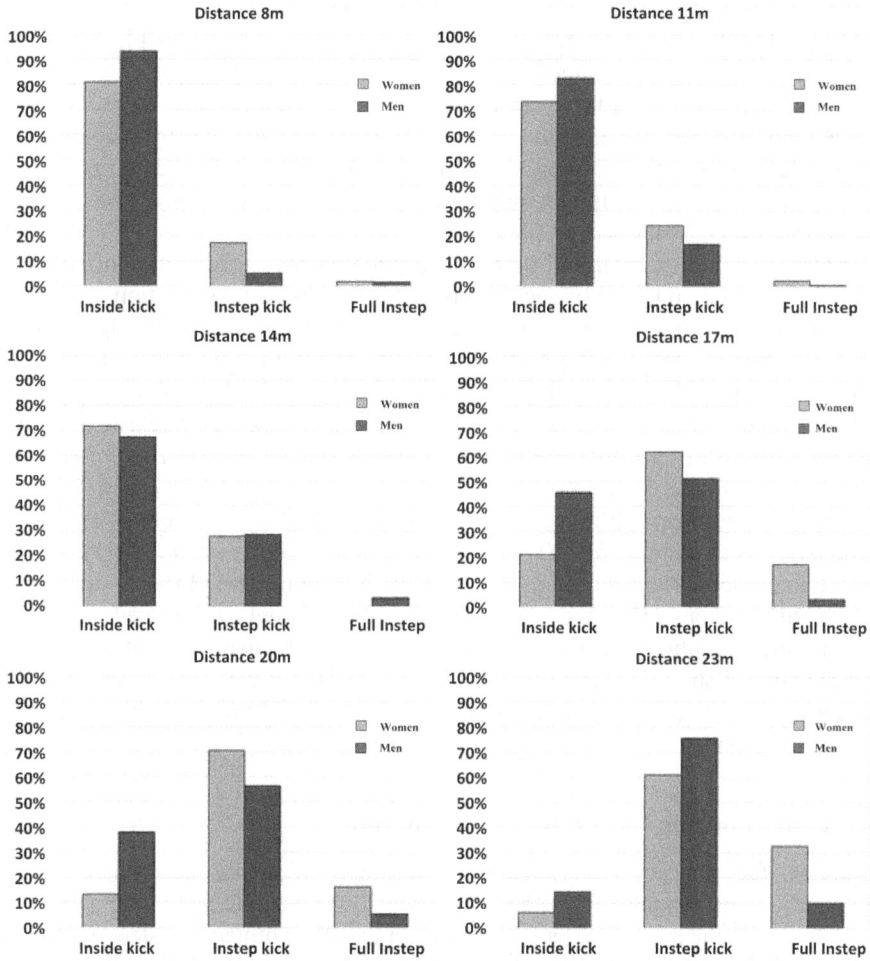

Figure 10.3 Kicking techniques used from six different distances

Weather conditions and the choice of footwear play an important role in soccer games. For most game situations the soccer shoe should provide adequate traction, preventing the player from slipping (Hennig, 2006; Hilgers & Walther, 2011; Sterzing *et al.*, 2007).

A study was conducted to analyse gender effects of different soccer shoe models on running time in a traction course. Twenty female and 20 male soccer players ran an agility traction course with four different shoe models. The functional traction course consisted of straight acceleration and slalom sections. Overall running time and three split times (first sprint, slalom, and second sprint) were recorded. Because soccer players adapt their movements to given ground-to-shoe interface conditions, running time results from obstacle courses can be employed to judge the traction properties of shoes (Hennig, 2006; Sterzing *et al.*, 2009b).

Due to their less powerful movements, it was expected that female players may benefit from different shoe models than the male players, because they do not need as much traction as the men. But no statistically significant interaction of gender and shoe design was found. Running time was not affected by shoe design but gender had a statistically highly significant ($p < 0.01$) influence on running time for all shoe conditions. Running times of the male soccer players was about 10 % faster than for the female players.

Providing good traction and therefore enhancing the performance of soccer players is just one purpose of soccer shoes. Another important function is injury protection (Lake, 2000). This means that a soccer shoe should provide high translational traction to allow fast acceleration and deceleration movements. However, a high rotational traction may increase the risk of knee injuries (Lambson *et al.*, 1996; Valiant, 1993). This is important for female soccer players because of their much higher risk of knee injuries. On average the risk of an ACL rupture is two to three times higher in female than in male soccer players. Additionally, female players get injured much more often in situations without any contact with an opponent player (Agel *et al.*, 2007; Dick *et al.*, 2007; Waldén *et al.*, 2011).

The four soccer shoes tested in this study were characterized by very different stud configurations. However, none of the shoe models influenced running performance significantly. It can be assumed that mechanical rotational traction varies between shoe models and hence the injury risk as well. Therefore, female soccer players should wear shoes showing low rotational traction without sacrificing performance.

5. Gender specific soccer shoe construction

Women's feet are not just scaled down versions of men's feet (Frey, 2000; Wunderlich & Cavanagh, 2001; Krauss *et al.*, 2008). Female feet – compared to male feet with the same foot length – have a more flat, pointed and slender shape whereas male feet are more voluminous (Krauss *et al.*, 2008). Especially in soccer, a tight fitting shoe is very important: it can improve ball handling, stability inside the shoe and traction on the field (Sterzing & Hennig, 2005; Sterzing *et al.*, 2011). Therefore, shoe fit and especially shoe lasts have to match the gender specific foot morphologies.

In addition to foot morphology data, it is necessary to know the wishes and demands that female and male soccer players expect from their footwear. A questionnaire was distributed to 300 female and 204 male soccer players (Althoff & Hennig, 2012; Althoff & Hennig, 2014). They were asked to rank 11 shoe features from most to least important. The results are shown in Figure 10.4. Comfort was ranked as the most important shoe feature by both genders and it was even more important for female players. Ball sensing or ball touch, stability inside the shoe and traction on the field were also desirable shoe properties. Here, the female players ranked stability as more important. Comfort as well as stability

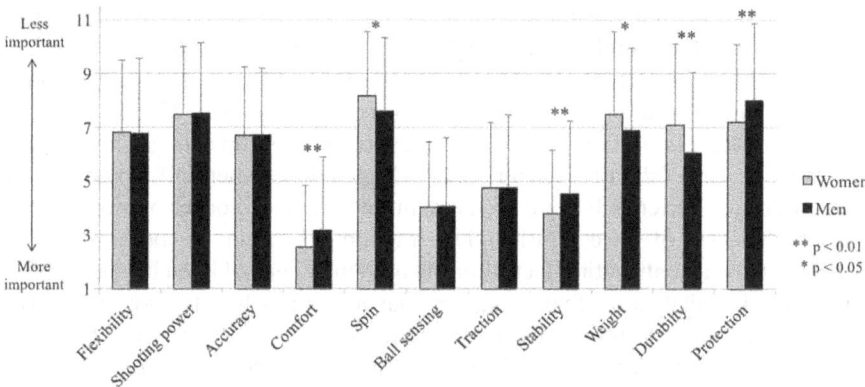

Figure 10.4 Priority ranking of soccer shoe properties (1 = most important, 11 = least important)

are closely related to shoe fit and both features are more important for female players. Injury protection was more important for the female players even though it was ranked as surprisingly unimportant by both genders when compared to performance-enhancing features.

Based on past experiences with their own shoes, soccer players were also asked to identify their preferences of shoe width, stud length, hardness of shoe upper and flexibility of the outsole on a five-point scale. Table 10.1 shows the results of desired shoe width. Depending on foot region, one third to one half of all soccer players were not satisfied with shoe fit. More women than men would like to have a (slightly) narrower shoe fit. Largest gender differences were found in the midfoot region: More than one third of the female players (36.9 %) would prefer a (slightly) narrower fit whereas nearly half of the male players – who are not satisfied with shoe fit – would prefer a (slightly) narrower (18.4 %) and the other half a (slightly) wider (19.9 %) fit. These results underline the necessity of gender specific shoe fit.

Table 10.1 Preferred changes in shoe fit for the forefoot, midfoot and rearfoot regions (**p < 0.01).

		Narrower	*Slightly narrower*	*No change*	*Slightly wider*	*Wider*
Forefoot	Women	7.4%	30.0%	51.2%	11.1%	0.3%
	Men	4.0%	26.2%	47.5%	20.3%	2.0%
Midfoot**	Women	3.7%	33.2%	52.5%	9.8%	0.7%
	Men	1.5%	16.9%	61.7%	18.9%	1.0%
Rearfoot**	Women	2.7%	26.7%	66.2%	4.4%	0.0%
	Men	3.5%	21.9%	65.2%	9.0%	0.5%

Further gender differences existed for flexibility of outsole and stud length. A (slightly) more flexible outsole would be preferred by 39% of the female soccer players and just 9% would prefer a (slightly) stiffer outsole. For the men, a higher percentage of soccer players would prefer a (slightly) stiffer outsole (17%) and a lower percentage would prefer a (slightly) more flexible outsole (28%).

Approximately half of the male soccer players, who would change stud length, would like to have (slightly) longer studs (15%) and the other half would like to have (slightly) shorter (18%) studs. As mentioned, female soccer players have a much higher risk of knee injuries. The traction behaviour of soccer shoes is suspected to be a contributing factor for the high incidence of knee injuries. Thus, it is surprising that more female soccer players would like to have (slightly) longer (23%) than (slightly) shorter (7%) studs. Either female soccer players are not aware of the higher injury risk by inadequate shoe constructions with long studs or they know about this risk, but focus more on performance rather than injury prevention.

6. Conclusions

The analysis of World Cup matches as well as studies with recreational soccer players under controlled conditions show that female soccer players kick the ball more often with the instep of the foot. Because of their lower muscle strength female players have to use different kicking techniques to cover same distances as male players. However, kicking techniques generating higher ball velocities are less accurate. Therefore, female soccer players should particularly practise kicking with the instep and inside to improve kicking accuracy and velocity. Furthermore, they could improve kicking performance by strengthening the muscles that stabilize the ankle joint during contact with the ball. The kicking accuracy studies show that female players kick less often with stable ankle joints and that these kicks are less accurate.

Regarding shoe construction, it is necessary that shoe fit matches the foot morphology of female and male soccer players. Comfort is by far the most important shoe feature and it is closely related to performance. Many female players are not satisfied with shoe fit and most of them would prefer a narrower shoe width. Nearly one quarter of the women players would also prefer longer studs. But female players – compared to their male counterparts – have a much higher knee injury risk and longer studs could further increase this risk. Therefore, women soccer shoes should be designed to reduce injury risks without sacrificing performance.

References

Agel, J., Evans, T. A., Dick, R., Putukian, M., & Marshall, S. W. (2007). Descriptive epidemiology of collegiate men's soccer injuries: National collegiate athletic association injury surveillance system, 1988–1989 through 2002–2003. *Journal of Athletic Training, 2*(42), 270–277.

Althoff, K. & Hennig, E. M. (2012). What female and male soccer players expect from their shoes. In *Proceedings of the 3rd World Conference on Science and Soccer*, p. 82.

Althoff, K. & Hennig, E. M. (2014). Criteria for gender-specific soccer shoe development. *Footwear Science*, 6(2), 89–96.

Althoff, K., Kroiher, J. & Hennig, E. M. (2010). A soccer game analysis of two World Cups: Playing behavior between elite female and male soccer players. *Footwear Science*, 2(1), 51–56.

Althoff, K., van Scherrenburg, K., Gürke, K., & Hennig, E. M. (2013). Entwicklungen im Frauenfußball – Eine Videoanalyse der Frauenfußball WM 2003 und 2011. In F. Mess, M. Gruber & A. Woll (Hrsg.), Sportwissenschaft grenzenlos?! (S. 71). Hamburg: Feldhaus Ed. Czwalina.

Andersson, H., Ekblom, B. & Krustrup, P. (2008). Elite football on artificial turf versus natural grass: Movement patterns, technical standards, and player impressions. *Journal of Sports Sciences*, 26(2), 113–122.

Åstrand, I., Åstrand, P.-O., Hallbäck, I., & Kilbom, A. (1973). Reduction in maximal oxygen uptake with age. *Journal of Applied Physiology*, 35(5), 649–654.

Bradley, P. S., Dellal, A., Mohr, M., Castellano, J. & Wilkie, A. (2014). Gender differences in match performance characteristics of soccer players competing in the UEFA Champions League. *Human Movement Science*, 33, 159–171.

Dick, R., Putukian, M., Agel, J., Evans, T. A., & Marshall, S.W. (2007). Descriptive epidemiology of collegiate women's soccer injuries: National Collegiate Athletic Association injury surveillance system, 1988–1989 through 2002–2003. *Journal of Athletic Training*, 42(2), 278–285.

Drinkwater, B. (1989). Das Training weiblicher Athleten. In A. Dirix, H. G. Knuttgen, & K. Tittel (Eds.), Olympia Buch der Sportmedizin (pp. 265–278). Köln: Deutscher Ärzte-Verlag.

FIFA. (2007). FIFA Big Count 2006. Zugriff am 23.06.2015 unter http://www.fifa.com/ mm/document/fifafacts/bcoffsurv/bigcount.statspackage_7024.pdf

FIFA. (2015). Women's football – Background information. Zugriff am 23.06.2015 unter http://resources.fifa.com/mm/document/footballdevelopment/women/02/60/99/71/fifa-background-paper_womensfootball_may2015_neutral.pdf

Frey, C. (2000). Foot health and shoewear for women. *Clinical Orthopaedics and Related Research, 372*, 32–44.

Frontera, W. R., Hughes, V. A., Lutz, K. J., & Evans, W. J. (1991). A cross-sectional study of muscle strength and mass in 45- to 78-yr-old men and women. *Journal of Applied Physiology, 71*(2).

Gokeler, A., Zantrop, T., & Jöllenbeck, T. (2010). Vorderes Kreuzband – Epidemiologie. *GOTS-Expertenmeeting*, 3–14.

Hennig, E. M. (2006). Biomechanische Methoden zur Optimierung und Evaluation von Fußballschuhen. *Orthopädieschuhtechnik, 6*, 20–24.

Hennig, E. M., Althoff, K., & Hömme, A. K. (2009). Soccer footwear and ball kicking accuracy. *Footwear Science*, 1(supl. 1), 85–87.

Hilgers, M. P. & Walther, M. (2011). Evolution of soccer shoe design. *International Journal of Athletic Therapy and Training, 16*(3), 1–4.

Hollmann, W. & Hettinger, T. (2000). Sportmedizin – Grundlagen für Arbeit, In *Training und Präventivmedizin*. Stuttgart, New York: Schattauer.

Horton, M. G. & Hall, T. L. (1989). Quadriceps femoris muscle angle: normal values and relationships with gender and selected measures. *Physical Therapy, 69*(11), 897–901.

König, K., Reindell, H., Musshoff, K., Roskamm, H., & Kessler, M. (1961). Das Herzvolumen und die körperliche Leistungsfähigkeit bei 20–60jährigen gesunden Männern. *Archiv für Kreislaufforschung, 35*(1–2), 37–67.

Krauss, I., Grau, S., Mauch, M., Maiwald, C., & Horstmann, T. (2008). Sex-related differences in foot shape. *Ergonomics, 51*(11), 1693–1709.

Krustrup, P., Mohr, M., Ellingsgaard, H., & Bangsbo, J. (2005). Physical demands during an elite female soccer game: Importance of training status. *Medicine and Science in Sports and Exercise, 37*(7), 1242–1248.

Lake, M. J. (2000). Determining the protective function of sports footwear. *Ergonomics, 43*(10), 1610–1621.

Lambson, R. B., Barnhill, B. S., & Higgins, R. W. (1996). Football cleat design and its effect on anterior cruciate ligament injuries: A three-year prospective study. *American Journal of Sports Medicine, 24*(2), 155–159.

Lindle, R. S., Metter, E. J., Lynch, N. A., & Fleg, J. vL. (1997). Age and gender comparisons of muscle strength in 654 women and men aged 20–93 yr. *Journal of Applied Physiology, 83*, 1581–1587.

Mohr, M., Krustrup, P., & Bangsbo, J. (2003). Match performance of high-standard soccer players with special reference to development of fatigue. *Journal of Sports Sciences, 21*(7), 519–528.

Neder, J. A., Nery, L. E., Shinzato, G. T., Andrade, M. S., Peres, C., & Silva, A. C. (1999). Reference values for concentric knee isokinetic strength and power in nonathletic men and women from 20 to 80 years old. *Journal of Orthopaedic and Sports Physical Therapy, 29*(2), 116–126.

O'Brien, M. (1995). Women and sport. *Sports Exercise and Injury, 1*, 131–137.

Roskamm, H., Reindell, H., Musshoff, K., & König, K. (1961). Die Beziehungen zwischen Herzgröße und Leistungsfähigkeit bei männlichen und weiblichen Sportlern im Vergleich zu männlichen und weiblichen Normalpersonen. *Archiv für Kreislaufforschung, 35*(1–2), 67–102.

Schmidt, R. F., Lang, F. & Heckmann, M. (2010). Physiologie des Menschen. Heidelberg: Springer.

Sterzing, T. & Hennig, E. M. (2005). Stability in soccer shoes – The relationship between perception of stability and biomechanical parameters. In T. Reilly, J. Cabri & D. Araújo (Eds.), *Science and Football V* (pp. 46–51), London: Routledge.

Sterzing, T., Hennig, E. M., & Milani, T. L. (2007). Biomechanische Anforderungen der Fußballschuhkonstruktion. *Orthopädie-Technik, 9*, 646–655.

Sterzing, T., Lange, J., Wächtler, T., Müller, C., & Milani, T. (2009a). Velocity and accuracy as performance criteria for three different soccer kicking techniques. In A. J. Harrison, R. Anderson & I. Kenny (Eds.) *Proceedings of the 27th International Society of Biomechanics in Sports*, pp. 243–246, Limerick: University of Limerick.

Sterzing, T., Müller, C., Hennig, E. M. & Milani, T. L. (2009b). Actual and perceived running performance in soccer shoes: A series of eight studies. *Footwear Science, 1*(1), 5–17.

Sterzing, T., Müller, C., Wächtler, T., & Milani, T. L. (2011). Shoe influence on actual and perceived ball handling performance in soccer. *Footwear Science, 3*(2), 97–105.

Thein, L. A. & Thein, J. M. (1996). The female athlete. *Journal of Orthopaedic and Sports Physical Therapy, 23*(2), 134–148.

Valiant, G. A. (1993). Friction – slipping – traction. *Sportverletzung Sportschaden, 7*(04), 171–178.

Waldén, M., Hägglund, M., Werner, J., & Ekstrand, J. (2011). The epidemiology of anterior cruciate ligament injury in football (soccer): a review of the literature from a gender-related perspective. *Knee Surgery, Sports Traumatology, Arthroscopy, 19*(1), 3–10.

Weineck, J. (2010). Sportbiologie (10. überarbeitete und erweiterte. Aufl.). Balingen: Spitta.

Wilmore, J. H. & Costill, D. L. (1994). *Physiology of sport and exercise*. Champaign: Human Kinetics.

Wunderlich, R. E. & Cavanagh, P. R. (2001). Gender differences in adult foot shape: implications for shoe design. *Medicine and Science in Sports and Exercise, 33*(4), 605–611.

11 Plantar pressure distribution patterns and a one-year aging process during soccer specific movements on a grass surface

Eric Eils

1. Introduction

Injuries are a common problem in elite and recreational soccer players (Junge *et al.*, 2002) and overuse injuries, which make up to 28% of all injuries in elite soccer players (Ekstrand *et al.*, 2011), play an important role in this context. Different extrinsic risk factors including repetitive impacts, surface properties, movement types and shoe properties may be involved in the etiology of overuse injuries (Dvorak *et al.*, 2000; Inklaar, 1994; Knapp *et al.*, 1998; Kristenson *et al.*, 2015; O'Kane *et al.*, 2015). To investigate the influence of these risk factors, it is essential to gain quantitative information concerning the foot-loading character-istics during soccer specific movements. Knowledge about the amount and exact location of the load, acting on the foot sole, is important to understand the etiol-ogy of injuries and may also help to develop specific foot/shoe/insole designs to prevent overuse injuries. During the last decade, several studies have analysed the foot–shoe–ground interaction during soccer specific movements on different surfaces with comparable results, discovering movement specific high loading patterns and areas under the sole of the foot (Eils & Streyl, 2005; Eils *et al.*, 2004; Ford *et al.*, 2006; Orendurff *et al.*, 2008; Queen *et al.*, 2008; Sims *et al.*, 2008; Wong *et al.*, 2007). Changes in the damping capacity of the foot/shoe/insole complex have rarely been evaluated, but it has been demonstrated that an aging process of six weeks leads to an increase in peak pressures in a soccer shoe (Eils *et al.*, 2001). However, a six-week aging process does not reflect the real situation when soccer shoes will be used over a much longer period. It can be assumed that the reduced damping capacity of a soccer shoe over time in combination with excessive loading of the foot in soccer may be a risk factor for the etiology of overuse injuries. Therefore, it is of interest how the damping capacity changes over a longer period of time and if exchanging the insole or the total shoe/insole complex can reduce the loading of risk bound foot areas.

The following chapter will focus on pressure distribution measurements during soccer specific movements. First, an analysis of typical plantar foot loading areas during four different soccer specific movements on grass is presented. Then, pres-sure distribution analyses of a one-year aging process on the shoe/insole complex during the same four soccer specific movements will be presented and discussed in the context of the current literature.

2. Methods

Twenty-one adult experienced male soccer players (mean age: 26±2 years, mean mass: 79±5 kg, mean height: 183±6 cm) participated in the first study to identify characteristic pressure distribution patterns during soccer specific movements (Eils *et al.*, 2004). Informed consent was obtained from all of them. Eleven players (mean age: 25±1 years, mean mass: 80±7 kg, mean height: 183±5 cm) participated also in the second study to investigate the effect of a one-year shoe/insole aging process from regular use on plantar pressures (Eils & Streyl, 2005). All players were free of injuries at the start and time of the studies and played soccer on a regular basis on a moderate performance level. Subjects wore a soccer shoe with a 12-stud traditional (elliptic) plate (Air Zoom Brasilia F.G., Nike, Inc.) in appropriate shoe sizes.

The Pedar Mobile system (Novel GmbH, Munich) was used to collect plantar pressure distribution patterns. Four typical soccer specific movements (normal run, sidecut, sprint, kick – the supporting leg) were analysed on grass surface on a soccer field. The one-year follow-up measurements were also completed on grass on the same field and at the same spot and similar surface conditions concerning density and moisture.

The normal run was performed at a velocity of 4.2 m/s. Subjects had to run a distance of several metres to reach the appropriate velocity and measurements were made within the last 25 m out of 150 m. The cutting manoeuvres were performed at approximately 70% of maximum speed within a slalom course that consisted of three cones on each side at equal distances of 12 m (Figure 11.1). Subjects ran from cone to cone and performed a sidecut with the outer leg at each cone. The sprint was performed at maximum speed over a distance of 40 m. The velocities for normal run, cut and sprint were controlled by timing lights. Subjects performed five individual steps in an approach before shooting the ball when analysing the supporting leg in goal shooting. All soccer specific movements were repeated three to five times to obtain a sufficient number of measurements for further analysis (finally nine steps for the normal run, cutting manoeuvre, and sprint; five steps for the goal shot).

For the one-year follow-up, the same procedures were repeated for three different shoe conditions in balanced order: (i) one-year used shoe and insole, (ii) one-year used shoe and new insole, (iii) new shoe and new insole.

The pressure distribution pattern of each step was subdivided into 10 different areas using a standardized mask. The areas were the medial and lateral heel, the medial and lateral midfoot, the medial, central and lateral forefoot, the hallux, the second toe and the lateral toes. For these, relative loads and peak pressures were measured and calculated for each step. Relative loads represent the local impulses divided by the sum of all local impulses, taking into account both magnitude and duration of loading. Both pressure variables are used to characterize foot loading. Peak pressures represent the maximal loading of a single pressure sensor across the underlying anatomical structures. They are used to describe the maximum loading of specific foot structures. The mean was calculated for each of the above mentioned foot areas and both parameters (relative loads, peak pressures).

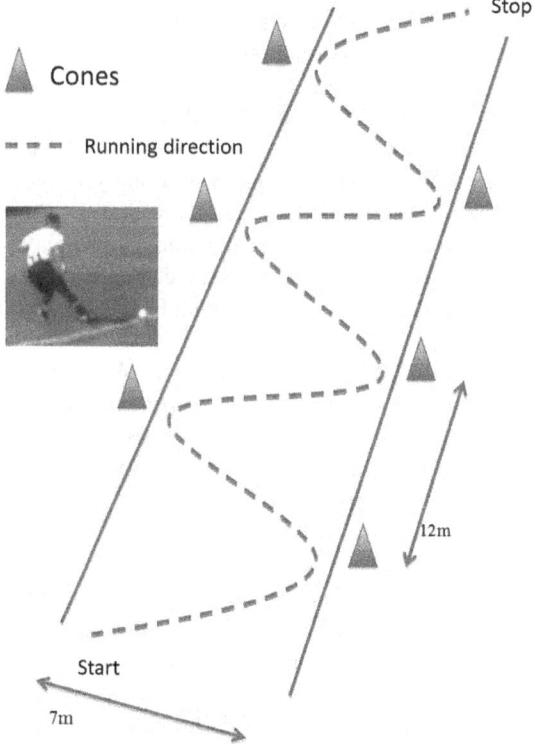

Figure 11.1 The experimental set-up (7 × 40 m) for the sidecut and the 30 m sprint (straight without cones)

A repeated measures ANOVA using the within factors movement (run, cut, sprint, kick) was used to focus on changes in foot loading characteristics. Due to the limited number of eleven participants within the second part of the investigation, the non-parametric Friedman test (used shoe and insole, used shoe and new insole, new shoe and insole) was used for the evaluation of the one-year aging process. The alpha-level for both analyses was set to 5% and adjusted accordingly due to multiple testing. For post-hoc comparisons, the Bonferroni-Holm test and the non-parametric Wilcoxon test were used, respectively.

3. Results

Table 11.1 shows the results in foot loading characteristics and Table 11.2 the results for the one-year aging process. The structure of Table 11.2 is different compared to Table 11.1 to enable a clearly arranged comparison between aging conditions.

The factor movement has a significant effect on relative loads and peak pressures in all areas (last column in Table 11.1). Post-hoc evaluation revealed characteristic

Table 11.1 Parameters of pressure distribution during four different movement conditions.

Peak Pressure [kPa]

	Run	Cut	Sprint	Kick	Statistics
med heel	284±50	703±143[c]	60±26[c]	707±129[c]	$F(3,60) = 2\,54$, $p<0.001$; eta^2=.93
lat heel	279±49	516±95 [c]	58±24[c]	741±163[c]	$F(3,60) = 215$, $p<0.001$; eta^2 = .92
med midfoot	137±30	235±64[c]	58±29[c]	285±114[c]	$F(3,60) = 64$, $P<0.001$; eta^2 = .76
lat midfoot	184±30	164±48	94±26[c]	389±108[c]	$F(3,60) = 1\,04$, $p<0.001$; eta^2 = .84
med forefoot	405±92	688±123[c]	589±163[c]	282±105[c]	$F(3,60) = 72$, $p<0.001$; eta^2 = .78
cent forefoot	323±66	322±80	397±71[c]	299±75	$F(3,60) = 11$, $P<0.001$; eta^2 = .35
lat forefoot	277±36	126±33[c]	289±61	404±101[c]	$F(3,60) = 77$, $P<0.001$; eta^2 = .80
hallux	353±77	504±107[c]	474±112[c]	381±89	$F(3,60) = 19$, $p<0.001$; eta^2 = .49
sec toe	189±47	243±64[c]	249±72[c]	266±65[c]	$F(3,60) = 12$, $p<0.001$; eta^2 =.38
lat toes	197±67	187±63	222±45	278±76[c]	$F(3,60) = 16$, $p<0.001$; eta^2 =.45

Relative loads [%]

	Run	Cut	Sprint	Kick	Statistics
med heel	8.2±2.4	17.5±3.7[c]	1.7±1.0[c]	9.5±2.5[b]	$F(3,60) = 236$, $p<0.001$; eta^2 = .92
lat heel	9.1±2.4	11.7±3.2[c]	1.9±1.2[c]	13.8±3.5[c]	$F(3,60) = 129$, $p<0.001$; eta^2 = .87
med midfoot	2.7±1.2	5.9±1.8[c]	0.6±0.5[c]	2.8±1.2	$F(3,60) = 82$, $P<0.001$; eta^2 = .81
lat midfoot	10.3±1.9	3.2±1.4[c]	3.7±1.1[c]	13.6±3.3[c]	$F(3,60) = 159$, $p<0.001$; eta^2 = .89
med forefoot	19.6±3.4	27.7±4.1[c]	29.4±5[c]	11.0±5.1[c]	$F(3,60) = 133$, $p<0.001$; eta^2 = .87
cent forefoot	16.0±2.5	10.0±2.0[c]	20.4±2.5[c]	11.6±3.2[c]	$F(3,60) = 129$, $P<0.001$; eta^2 = .87
lat forefoot	18.3±1.7	5.3±1.6[c]	18.4±4.2	20.4±2.5[b]	$F(3,60) = 164$, $P<0.001$; eta^2 = .89
hallux	10.3±2.2	13.7±3.8[c]	15.4±3.1[c]	9.1±2.1[a]	$F(3,60) = 61$, $p<0.001$; eta^2 = .75
sec toe	5.1±1.7	5.4±1.5	8.5±2.6	5.8±1.5	$F(3,60) = 29$, $p<0.001$; eta^2 = .60
lat toes	4.3±1.8	33±1.7[b]	5.7±1.5[b]	5.0±1.8	$F(3,60) = 16$, $p<0.001$; eta^2 =.44

a, b, c indicate significant differences in movement compared to the normal run with an alpha-level of 0.05, 0.01 and 0.001, respectively.

patterns for cutting, sprinting and kicking compared to running. In running, the areas with the highest peak pressures are the heel, the forefoot (with reduction from medial to lateral), and the hallux. In cutting, the heel, the medial forefoot and the hallux, in sprinting, the forefoot and the toes (with a reduction from medial to lateral), and in kicking (supporting leg) the heel, the lateral mid- and forefoot. In some areas, these changes are particularly high (e.g. an increase of 250% and 170% for the medial heel and forefoot in cutting, an increase of 145% for the first ray in sprinting, and 270% for the lateral heel in kicking) (Table 11.1). An analysis of the load shift between conditions (cutting, sprinting, kicking) and normal run reveals typical loading patterns: there is a significant shift to the heel, medial mid- and forefoot, and hallux in cutting, a shift to the first and second ray and the lateral toes in sprinting, and a shift to the heel and the lateral part of the foot in kicking (supporting leg). Figure 11.2 shows the significant load shifts for each condition when compared to running.

The comparison of the three aging conditions shows inconsistent results for peak pressures as well as relative loads (Table 11.2). Focusing on peak pressures, there is the tendency towards pressure reduction in some areas when replacing the used insole with a new one or additionally exchanging the shoe. For example, reduced values result for the areas of the heel, medial mid- and forefoot, and hallux in running, in cutting for the areas of the heel, midfoot, medial forefoot, and hallux, in sprinting the forefoot and hallux areas, and in kicking the heel, lateral mid- and forefoot, second toe and lateral toes. There are some areas that even show an opposite or mixed tendency, e.g. the central forefoot in running and cutting, the heel in sprinting or the medial and central forefoot in kicking. However, a significant pressure reduction exists only for the medial midfoot in running, sprinting, and kicking. Some of the changes are only due to the new dampening insole, some due to the whole complex of new insole and shoe.

The parameter relative loads also shows inconsistent results and statistically significant changes exist for the midfoot area in cutting and the second toe in kicking.

Relative loads

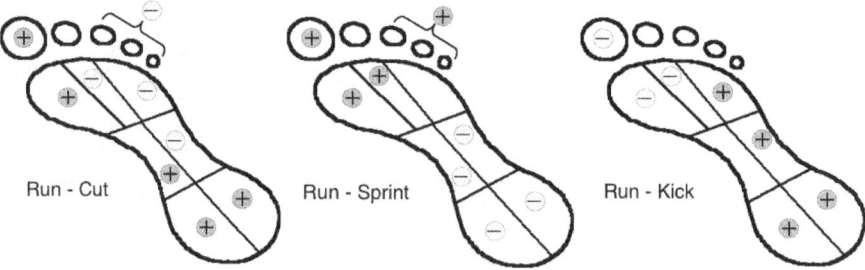

Figure 11.2 Significant load shifts for each condition compared to running. A '+' indicates an increase and a '−' a decrease in relative loads compared to running

Table 11.2 Parameters of pressure distribution during four different movement conditions (run, sidecut, sprint, kick) and three different aging conditions (1Y=1 year old shoe and insole, NI = 1 year old shoe and new insole, NS = new shoe and new insole). Modified with permission (Eils & Streyl, 2005).

Peak Pressures [kPa]		heel		midfoot		forefoot			toes		
		med heel	lat heel	med midfoot	lat midfoot	med forefoot	cent forefoot	lat forefoot	hallux	sec toe	lat toes
Run	1Y	267±90	255±67	118±36	158±40	336±158	264±94	239±49	296±128	129±69	155±99
	NI	263±60	257±50	108±33	166±51	332±149	272±97	248±66	273±106	138±77	158±91
	NS	255±64	250±60	93±29a	145±46	330±144	277±107	240±69	286±85	131±71	137±93
Cut	1Y	706±271	491±219	228±89	119±40	566±206	277±78	101±25	381±117	168±65	119±54
	NI	703±248	446±187	186±62	91±30	513±159	307±69	111±37	333±84	183±62	127±58
	NS	635±268	459±175	160±57a	92±40	536±130	298±92	107±51	364±110	185±75	132±84
Sprint	1Y	31±23	32±23	53±26	77±26	542±250	341±102	253±49	412±208	181±84	171±84
	NI	45±28	42±28	46±24	71±19	546±80	349±80	253±37	367±121	191±72	187±103
	NS	51±3	47±31	45±24	72±18	516±217	340±135	245±52	386±112	178±91	173±104
Kick	1Y	628±126	665±123	281±106	355±77	201±85	216±61	343±103	303±175	196±85	208±101
	NI	584±138	668±171	196±87a	312±66	241±135	231±72	334±88	305±112	173±82	199±77
	NS	592±86	555±187	186±68a	260±103	215±78	222±68	303±115	281±125	165±87	204±92
Relative loads [%]											
Run	1Y	8.4±2.6	9.5±2.3	4.3±1	10.8±1.2	19.3±2.8	16±2.1	19.6±3.4	7.8±2.9	4.1±1.7	3.9±1.8
	NI	8.9±3.1	10.2±2.1	4.1±1.4	10.7±2	19.9±3.2	16.6±2.1	19.3±3.8	7.1±2.4	3.2±1.8	3.8±2.2
	NS	9.4±3.2	9.4±2.3	3.1±1.3	10.3±1.3	20.8±3	16.6±2.2	18.9±2.5	8.3±1.9	3.4±1.5	3.7±2.3
Cut	1Y	20.4±5.7	11±4	7.2±1.8	3.4±1.1	29±6.1	11.1±3.1	5.8±1.8	10±2.9	3.9±1.4	2.4±1.1
	NI	21.5±4.7	11.5±4.2	5.8±2.2	2.5±1.1a	29.1±7	12.1±3.1	6.1±1.8	9.5±2.5	3.7±1.4	2.5±1.4
	NS	21.0±3.9	11.5±4.4	5.1±1.7b	2.8±0.9	30±6.2	11.4±2.5	5.6±1.8	10.4±1.8	3.8±1.9	2.5±1.9
Sprint	1Y	0.8±0.9	0.9±1	2.3±1.7	4.2±1.8	31.9±4.9	21.5±3	21±5.7	11.9±3.6	6.4±2.7	5±1.8
	NI	1.3±0.8	1.5±1.4	1.3±1.4	3.4±1.7	32.9±5.8	21.8±2.8	20.4±4.2	12.3±3.2	6.1±2.6	5±2.5
	NS	1.5±1	1.5±1.4	1.3±1.1	4.2±0.8	32.3±4.5	21.3±2.5	20.1±4	13.1±2.2	5.6±2.2	4.7±2.7
Kick	1Y	11.6±3.2	15.9±4.7	4.3±2.4	13.3±2.7	10.2±4.1	11±3.5	20.7±2.5	6.6±2.4	4.6±2.2	4.3±1.7
	NI	11.3±4.4	17.2±3.6	3.1±1.4	12.5±3.1	12±4.7	12±3	20.3±3.1	6.6±2.1	3.5±2a	3.9±1.3
	NS	13.8±8.3	16±5.7	3.7±3.2	12.7±5.4	11.6±4.6	11.3±2.4	19.3±4.6	6.1±2.4	3.4±1.4b	4.4±2.2

a or b indicate significant differences to 1Y-condition with an alpha-level of 0.05 or 0.01, respectively.

4. Discussion

This chapter has focused on pressure distribution analyses inside a soccer shoe to identify typical loading patterns of the plantar foot and to investigate a long-time aging process of the foot/insole complex during four soccer specific movements. The results revealed characteristic pressure distribution patterns with typical loading areas of the foot that are clearly related to the evaluated movements. In addition, the results show that the use of a new insole or a new shoe/insole complex after a one-year aging process of the material does not lead to anticipated load reductions within the typical loading areas of the foot.

Soccer specific movements lead to typical loading patterns of the sole of the foot. During the last decade, these results have been confirmed by other investigations (Ford *et al.*, 2006; Orendurff *et al.*, 2008; Queen *et al.*, 2008; Sims *et al.*, 2008; Wong *et al.*, 2007). Ford *et al.* (2006) investigated the foot loading patterns on natural grass and synthetic turf during cutting movements. The methods were similar to the ones presented above, and the results were also comparable, i.e. there was an increase in peak pressures and relative loads especially for the medial part of the heel and forefoot and the hallux. The amount of pressure is also high compared to normal running and underlines the potential for overloading these foot areas. Wong *et al.* (2007) investigated in-shoe plantar pressure distributions during running, cutting, and jump landing in soccer on an artificial turf surface. Again, methods and results generally agreed with the findings of the present investigation. The authors reported a significant increase of peak pressures and pressure-time integrals for the medial part of the foot in cutting. Finally, Orendurff and colleagues (2008) confirmed these results when evaluating foot pressures during running, cutting, jumping and landing. Therefore, it can be assumed that soccer specific movements performed in a soccer shoe lead to characteristic loading patterns of the foot sole with an enormous increase in peak loading of specific areas of the foot. These findings seem to be independent from the tested surface because the characteristic pressure distribution patterns occur on natural grass (Eils *et al.*, 2004; Ford *et al.*, 2006), artificial turf (Ford *et al.*, 2006; Orendurff *et al.*, 2008; Wong *et al.*, 2007), and red cinder (Eils *et al.*, 2004). At this point it has to be considered that the amount in pressure loading may still vary to some degree between surfaces as reported by Ford *et al.* (2006).

Although there is no absolute threshold for the development of overuse injuries, all these investigations indicate a potential risk for overloading specific foot areas reporting peak pressure values more than twice as high as the normal run in some areas of the foot. Certainly, other important factors like duration of exposure or frequency of loading play a role in the etiology of overuse injuries apart from just maximum values. From that point of view, attention should be paid to the frequent movements of sprinting and cutting and their loading on the medial side of the foot. On the other hand, the bony structures of the medial part of the foot are more robust to loading than lateral parts (e.g. the fifth metatarsal bone).

Thus, less frequent movements like kicking but with an extreme loading of the lateral aspect of the foot may also predispose some bony structures to injuries. Indeed, the majority of fifth metatarsal fractures are stress-fractures and mainly occur among young players during the preseason high-intensity training period. In general, fifth metatarsal fractures are rare but the stress nature of the injury might explain the high frequency of healing problems (Ekstrand & van Dijk, 2013). Finally, it has to be considered that other types of stresses such as bending, torsional and shear loading may be of importance when focusing on overuse or hard and soft tissue injuries. With traditional pressure insole devices these stress variables cannot be measured. Therefore, the use of new systems appears promising (Stief & Peikenkamp, 2015).

A more pronounced peak pressure increase as well as load shifts had been anticipated for the results from the shoes, being exposed to the one-year aging process, because a clear aging effect was reported in a previous investigation after an aging period of six weeks (Eils *et al.*, 2001). A significant increase of peak pressures above 10% has been found for the characteristic pressure areas during soccer specific movements, e.g. under the heel, mid- and forefoot area for the normal run, under the medial structures of the foot during cutting, under the forefoot in sprinting and under the lateral midfoot of the supporting leg in kicking. It remains unclear why a brief aging process seems to be detectable but not a long-time aging process. There are clear deterioration effects visible within the dampening structures of the insoles as well as the shoes for all subjects: the heel air cushion was broken and the dampening capability under the first metatarsal and the hallux was gone and reduced for most other areas. Figure 11.3 shows an example of a one-year used insole and shoe.

The soccer shoe also shows considerable changes, i.e. the used shoe appears to be shorter, wider and the sole is more bent (Figure 11.3, middle). It may be hypothesized that this could be the result of different aging processes taking place for the insole and the shoe in a sequential order: first, the insole is the primary damping element that is altered within the first weeks and thus leads to an increase in loading of specific foot structures; second the flexibility of the shoe (outer sole and material) increases over time and the foot is able to align to its individual form (e.g. flattening) thus resulting in an improved pressure distribution pattern over the whole foot sole leading to a reduction of peak pressures. It also has to be considered that players may adopt their motion pattern in relation to the external loading of the foot over time and these impact signals lead to a recalibration of motion control (Nigg, 2001). Due to the fact that the normal soccer shoe is rigid, it has a limited dampening capacity and the loading of the foot in soccer is considerably increased compared to running. It seems conceivable that an effective load reduction is not only performed by using new damping insoles but also by a modification of the movement. Players may adapt to bad footwear by modifying their movements in order to protect their body. However, this may often result in worse performance on the field because the players have to concentrate on body protection rather than the demands of the game.

Figure 11.3 A comparison of a new and one year old soccer shoes and insoles

5. Conclusion

This chapter has focused on plantar pressure distribution measurements inside a soccer shoe. The results demonstrated characteristic loading patterns of the foot during soccer specific movements. Some areas are exposed to excessive loading values, suggesting an increased risk for the development of overuse injuries. The evaluation of a one-year aging process of the shoe/insole complex revealed a non-linear aging process that may be influenced by different factors like material properties, shoe shape changes and the interaction between excessive loading and adaptation processes in motion patterns.

References

Dvorak, J., Junge, A., Chomiak, J., Graf-Baumann, T., Peterson, L., Rosch, D., & Hodgson, R. (2000). Risk factor analysis for injuries in football players. Possibilities for a prevention program. *The American Journal of Sports Medicine*. 28, S69–74.

Eils, E. & Streyl, M. (2005). A one year aging process of a soccer shoe does not increase plantar loading of the foot during soccer specific movements (in German). *Sportverletz Sportschaden*, 19, 140–145.

Eils, E., Streyl, M., Linnenbecker, S., Thorwesten, L., Völker, K., & Rosenbaum, D. (2001). Plantar pressure measurements in a soccer shoe: Characterization of soccer specific movements and effects after six weeks of aging. In E. Hennig & A. Stacoff (Eds.), Proceedings of 5th Symposium on Footwear Biomechanics (pp. 32–33), Schlieren: ETH Zürich.

Eils, E., Streyl, M., Linnenbecker, S., Thorwesten, L., Völker, K., & Rosenbaum, D. (2004). Characteristic plantar pressure distribution patterns during soccer-specific movements. *The American Journal of Sports Medicine*, 32, 140–145.

Ekstrand, J., Hagglund, M., & Walden, M. (2011). Injury incidence and injury patterns in professional football: The UEFA injury study. *British Journal of Sports Medicine*, 45, 553–558.

Ekstrand, J. & van Dijk, C. N. (2013) Fifth metatarsal fractures among male professional footballers: a potential career-ending disease. *British Journal of Sports Medicine*, 47, 754–758.

Ford, K. R., Manson, N. A., Evans, B. J., Myer, G. D., Gwin, R. C., Heidt, R. S., Jr. & Hewett, T. E. (2006). Comparison of in-shoe foot loading patterns on natural grass and synthetic turf. *Journal of Science and Medicine in Sport*, 9, 433–40.

Inklaar, H. (1994). Soccer injuries. II: Aetiology and prevention. *Sports Medicine*, 18, 81–93.

Junge, A., Rosch, D., Peterson, L., Graf-Baumann, T. & Dvorak, J. (2002). Prevention of soccer injuries: A prospective intervention study in youth amateur players. *The American Journal of Sports Medicine*, 30, 652–659.

Knapp, T. P., Mandelbaum, B. R. & Garrett, W. E. (1998). Why are stress injuries so common in the soccer player? *Clinics Sports Medicine*, 17, 835–853.

Kristenson, K., Bjorneboe, J., Walden, M., Ekstrand, J., Andersen, T. E. & Hagglund, M. (2016). No association between surface shifts and time-loss overuse injury risk in male professional football. *Journal of Science and Medicine in Sport*, 19, 218–221.

Nigg, B. M. (2001). The role of impact forces and foot pronation: A new paradigm. *Clinics in Sports Medicine*, 11, 2–9.

O'Kane, J. W., Gray, K. E., Levy, M. R., Neradilek, M., Tencer, A. F., Polissar, N. L. & Schiff, M. A. (2016). Shoe and field surface risk factors for acute lower extremity injuries among female youth soccer players. *Clinics in Sports Medicine*, 26(3), 245–250.

Orendurff, M. S., Rohr, E. S., Segal, A. D., Medley, J. W., Green, J. R., 3rd, & Kadel, N. J. (2008). Regional foot pressure during running, cutting, jumping, and landing. *The American Journal of Sports Medicine*, 36, 566–571.

Queen, R. M., Charnock, B. L., Garrett, W. E., Jr., Hardaker, W. M., Sims, E. L. & Moorman, C. T., 3rd. (2008). A comparison of cleat types during two football-specific tasks on FieldTurf. *British Journal of Sports Medicine*, 42, 278–284.

Sims, E. L., Hardaker, W. M. & Queen, R. M. (2008). Gender differences in plantar loading during three soccer-specific tasks. *British Journal of Sports Medicine*, 42, 272–277.

Stief, T. & Peikenkamp, K. (2015). A new insole measurement system to detect bending and torsional moments at the human foot during footwear condition: a technical report. *Journal of Foot and Ankle Reserach*, 8, 49. doi: 10.1186/s13047-015-0105-6.

Wong, P. L., Chamari, K., Mao de, W., Wisloff, U. & Hong, Y. (2007). Higher plantar pressure on the medial side in four soccer-related movements. *British Journal of Sports Medicine*, 41, 93–100.

12 Influence of traction on running performance and lower extremity loading of soccer players

Interplay of shoes, surfaces and training

Thorsten Sterzing

0. Overview

In soccer optimal traction of players is essential for game success as it translates directly into athletic performance. Anecdotally, superior grip of their soccer shoes gave the German team a performance margin over the Hungarian team in the 1954 Soccer World Cup final, contributing to its historic victory. The famous match showed that, remarkably, soccer shoes can influence the course of a game. Thus, it is regarded as the root of systematic soccer shoe innovation. Thereafter, research, design, development, and manufacturing of soccer shoes received increased attention, not only addressing athletic performance but also aspects of preventing injury risks. Subsequent research strongly focussed, and still does, on soccer shoe traction. Importantly, traction of players is always dependent on the interaction of the shoe with the respective surface played on (Sterzing *et al.*, 2009; Sterzing *et al.*, 2010; De Clercq *et al.*, 2014; Schrier *et al.*, 2014). Thereby, traction is assessed with specific regard to the various inherent game actions of soccer, for instance straight runs, cuts, turns, jumps, or foot plants preceding passes and kicks (Sterzing & Hennig, 2008; Sterzing *et al.*, 2009). Furthermore, players' bodily characteristics and their coordinative skill level also have to be considered during optimization of functional soccer traction (Sterzing, 2016).

This book chapter addresses current knowledge of traction effects on players' acceleration, cutting, and turning performance under consideration of their lower extremity loading. Specific attention is directed to the relationship of mechanical availability of traction and its biomechanical utilization, a key principle for optimizing soccer traction properties. It is further illustrated that utilized biomechanical traction during soccer specific actions is not only a function of the given shoe and surface characteristics, but also a function of players' anatomical, physiological and motor skill capabilities. These capabilities are based on rather stable resources, like players' anthropometrics and body composition, and on more easily adaptive resources, like neuro-muscular coordination assets. In this sense, adjustment to different traction environments and application of specific training regimens mark important additional factors when analysing traction properties directed towards performance enhancement and injury prevention of soccer players. This chapter addresses current knowledge of performance and injury prevention related to soccer traction. Furthermore, new pathways to increase traction functionality of soccer players are derived. Traction in soccer is dependent on individual player performance, the type of movement on the field, surface conditions and footwear construction.

1. Background

Soccer traction refers to the shear forces players generate during ground contact of game specific actions. Higher shear forces allow players to perform movements more explosively, whereas lower shear forces indicate inferior circumstances for athletic performance. Soccer traction properties are in particular essential during maximal accelerations and during rapid changes of direction, for instance cutting and turning movements that contain considerable braking and propulsion aspects.

In general, soccer traction properties are dependent on player characteristics, game specific actions, surface conditions, and shoes. However, when play is about to start the only adjustable factor to influence players' traction is the shoe, which explains the large interest in the functionality of soccer shoe outsole configurations. In a broader sense, soccer shoes aim to enhance comfort, performance, and injury prevention, known as the three major functional requirements of athletic footwear (Sterzing, 2015).

The functional design of a soccer shoe needs to respond to numerous and complex requirements, as shoe-to-foot, shoe-to-ground, and shoe-to-ball interaction have to be considered (Sterzing, 2015). Soccer shoe designs were already shown to be able to improve the quality of these interaction types (Hennig & Sterzing, 2010). It is noteworthy that official technical demands stated in FIFA Laws of the Game (2015–16) impose only minimal mandatory requirements regarding soccer shoe construction. Shoes are simply referred to as personal equipment, mandatory to wear, and to be safe. This rough framework provides plenty of room for manufacturers to create innovative shoe designs that aim to improve soccer shoe functionality. Thereby, the most important soccer shoe features to address are comfort, stability, traction, and ball sensing, as identified by a player questionnaire survey (Hennig & Sterzing, 2010). Among these features, traction has received the largest attention in scientific contexts, as it is a key performance indicator. Optimal traction is required during numerous soccer specific movements, in forward and backward orientation, in various directions, and performed at various speeds. As mentioned above, players' opportunities to provide themselves best traction circumstances are limited to their selection of shoes, once the more general playing environment, including surface type and weather conditions, is set. In this sense, a basic differentiation in the analysis of soccer traction should be made between natural clay, natural grass and artificial grass surfaces. Either type asks for specific shoe outsole constructions, which should be additionally differentiated regarding temporary weather effects and their general quality level. The influence of temporary weather conditions on shoe–surface interaction marks a basic factor to be differentiated regarding natural clay and grass surfaces. It is noteworthy that natural surfaces are strongly affected by different weather conditions, whereas artificial grass surfaces seem to be affected only to minor extent.

In the following, soccer traction properties are addressed solely by analysing shoe-to-ground interaction, while the shoe and the foot are regarded as one stable functional unit. Key soccer shoe components of the outsole configuration are the outsole itself and the stud configuration. Both of them can be systematically altered, while the variation of stud configuration has received larger attention. Stud configurations can be varied regarding their stud number, stud positioning, stud geometry, stud orientation, stud length, stud dimension/area, stud hardness, stud material, and others. Related functional effects on players' performance and loading appear not always obvious and require systematic evaluation. For the general analysis of athletic footwear properties, comprehensive evaluation techniques should be applied (Sterzing *et al.*, 2012). Thereby, detailed knowledge of the mechanical soccer shoe properties, their biomechanical effects on players, subsequent effects on players' athletic performance, and effects on players' subjective perception are retrieved, allowing systematic improvement of the shoe's functional design. In addition, adequately administered computer simulation techniques help to anticipate traction benefits of altered stud configurations. This can considerably minimize time and costs during soccer shoe design, development, and manufacturing processes, and should accompany respective functional evaluation processes.

2. Running performance

Competitive soccer has numerous performance demands, among them running performance is regarded a key aspect. Surface conditions and shoe outsole configurations alter the specific functionality of shoe-to-surface interaction, thereby influencing running performance (Krahenbuhl, 1974). It is noteworthy that the increase or decrease of running performance due to altered shoe-to-surface interaction is well perceivable by soccer players (Sterzing *et al.*, 2009). This observation confirms that psychological aspects also need to be considered during the examination and subsequent improvement of soccer traction properties.

The main traction components of soccer shoes are their outsole stiffness and their stud configuration. However, investigations of shoe outsole stiffness effects have rather been directed towards track and field running and sprinting. Theoretical considerations and certain research findings suggest stiffer soles should favour acceleration and sprinting performance of athletes, however summed research findings appear still inconclusive. Thus, it is suggested that shoe stiffness levels should be optimized for individual anthropometric and neuro-muscular characteristics of athletes (Stefanyshyn & Fusco, 2004; Toon *et al.*, 2009; Smith *et al.*, 2016). Effects of soccer shoe outsole stiffness to increase acceleration and agility running performance have not been systematically examined. Instead, soccer shoe stud configurations are the focus of attention and were shown to especially alter the agility running performance of players. It was illustrated that players are faster when wearing the right shoe for a given surface condition, and players are slower when wearing shoes that are less suited to them (Sterzing *et al.*, 2009). Thereby, stud configuration variation has bigger influence on running performance during more complex running tasks, incorporating changes of direction, than during comparably simple tasks, for instance short and straight acceleration. Typically, changes of direction feature a braking component, decelerating the player's movement in its initial direction, and a propulsion component, accelerating the player's movement towards the new direction. In contrast, straight acceleration or sprinting movements mainly feature propulsion components, while braking components appear to be minimal. The sequence of severe braking and strong propulsion during rapid changes of direction attributes more relevance to stud configuration characteristics for achieving best performance. During changes of direction high shear forces are generated at the shoe-to-surface interface in medio-lateral and in anterior-posterior orientation, often combined and referred to as resultant shear force. In particular the stud configuration of soccer shoes influence the magnitudes and loading rates of these shear forces, to large extent determining running performance of players.

For optimal running performance, a functional bonding of studs and surface is essential. Early on, typical stud configuration categories evolved, soft ground, firm ground, and hard ground, indicating their usage recommendation on natural grass surfaces. These were specifically developed in response to altered surface conditions due to different weather conditions (top of Figure 12.1).

Within these soccer shoe categories stud configurations appear to be further modified, for instance by different length of soft ground screw-in studs, by differently shaped firm ground studs, or by altered numbers of hard ground studs. Generally, such modifications are directed towards variation of stud number and positioning, of stud geometry and orientation, of stud length and dimension/area, of stud material and hardness, and other factors. Two different stud geometries

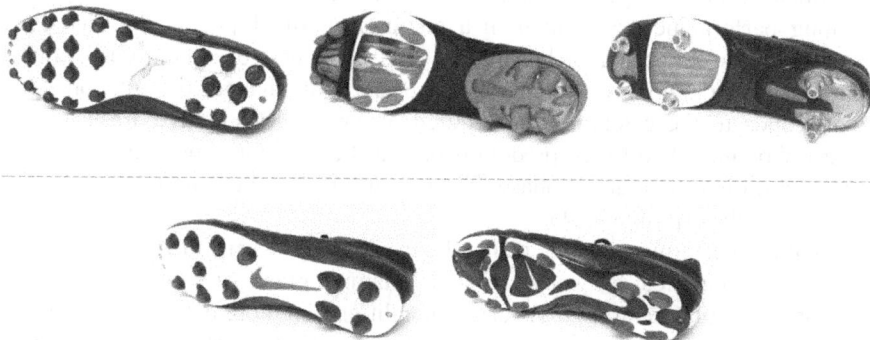

Figure 12.1 Hard ground (left), firm ground (middle), and soft ground (right) stud types, developed for altered natural grass surfaces due to climatic and temporal weather conditions (top); elliptic (left) and bladed (right) firm ground stud configurations of Nike Tiempo Premier™ (bottom)

are demonstrated by the elliptic and the bladed shapes of a firm ground natural grass stud configuration (bottom of Figure 12.1).

More recently, stud configurations were also adapted for usage on artificial grass surfaces. Baseline research revealed that among the traditional natural grass stud categories (top of Figure 12.1), soft ground studs appeared inadequate for playing on artificial grass surfaces (Sterzing *et al.*, 2010). These findings confirmed players' intuitive shoe selection for playing on artificial grass surfaces in times when research findings were not yet available (Müller *et al.*, 2009a). Firm ground natural grass stud configurations appeared suitable, while hard ground stud configurations were found best suited for playing on artificial grass surfaces. Based on these research findings, fine tuning of respective stud configurations took place and initial guidelines for specific artificial soccer turf stud configurations were derived. They state that appropriate stud configurations should feature a relatively high number of relatively short studs in order to achieve best running performance and provide reasonable lower extremity load management (Sterzing *et al.*, 2010).

Systematic differences in running performance were identified for stud length (up to 26%), stud geometry (up to 3%), stud type (up to 3%), and surface (up to 21%) (Sterzing *et al.*, 2009). In particular stud length had a large influence on running performance. It was confirmed that the range of stud lengths currently offered by manufacturers provides good grip to players (Müller *et al.*, 2009b; Sterzing *et al.*, 2009). As expected, removing studs completely resulted in severe loss of grip and decreased performance. However, reducing stud length by half induced a less severe decrease of running performance already, in contrast to shoes featuring regular stud length. Future research should target smaller variations of the current stud length range, as it is unclear whether less extensive modifications also have potential to enhance running performance of players. The increase of stud configurations' obstructive area has been identified as a basic principle to enhance agility running performance (Müller *et al.*, 2009b). Within functional limits, a larger obstructive area was demonstrated to increase agility running performance, as shown in a comparison of elliptic and bladed stud configurations. Primarily, the benefit of a larger obstructive area unfolds its function during the braking component when performing changes of direction. This corresponds to the comparison illustrated, where the obstructive areas of two soccer shoe models differed predominantly at the rearfoot region, an area more involved during the braking phase of cuts and turns than during the subsequent propulsion phase (Table 12.1).

Running performance also differed considerably when surface conditions varied, as quantified for snowy and icy natural grass surface conditions in contrast to regular natural grass surface conditions. In more detail, it was shown that a bladed firm ground stud configuration allowed better agility running performance than an elliptic firm ground stud configuration on the snowy and icy surface. This was probably the case because the bladed stud configuration allowed better penetration into the partly frozen surface condition examined (Sterzing *et al.*, 2009).

Table 12.1 Obstructive area size [cm²] at the lateral and medial edge of soccer shoes for examples of elliptic and bladed stud geometry (bottom of Figure 12.1), leading to a 3% advantage in agility running performance of the bladed stud geometry.

Stud geometry	Obstructive area medial [cm²]		Obstructive area lateral [cm²]	
	Rearfoot	Forefoot	Rearfoot	Forefoot
elliptic	3.6	4.6	3.6	6.2
bladed	4.9	4.8	4.9	6.4

Efforts to increase soccer agility running performance by improving traction characteristics of soccer shoes are ongoing. So far, research on the respective functionality of stud configurations provides rather general guidance on which characteristics to favour for given surface conditions. However, for most aspects of potential stud modification the exact beneficial thresholds of certain stud numbers, positioning patterns, geometries, orientations, dimensions, lengths, materials, and hardnesses still need to be determined. Another promising but more sophisticated approach to enhance soccer agility running performance is to combine the conceptual benefits observed for isolated modifications of stud features. Thus, research approaches targeting multifactorial optimization of the various stud characteristics should be applied to find out their optimal interaction effects for a given surface condition. Such approaches are suitable for combining the specific benefits of different stud characteristics, lifting soccer traction properties to advanced levels. Additionally, future research, design, and development approaches should include the effects of outsole stiffness. Thereby, general outsole stiffness but also segmented outsole stiffness should be considered. In due course, interaction effects of stud configuration and outsole stiffness should be investigated as it is likely that different stud configurations would benefit from specifically adjusted outsole stiffness characteristics.

3. Lower extremity loading

Incidental and repetitive loading of soccer players is linked to the occurrence of acute and overuse injuries (Fuller *et al.*, 2007a; Fuller *et al.*, 2007b). Shoe-to-surface interaction during soccer actions influences in particular lower extremity loading of players, also triggering specific alteration of lower leg kinematics (Müller *et al.*, 2010a; Schrier *et al.*, 2014). Loading magnitudes, loading rates, and corresponding joint moments provide a good estimate of the forces evoked during ground contacts of soccer specific movements. Variable scores are dependent on players' anthropometrics and their movement execution dynamics, but change due to the characteristics of shoe and surface conditions. Ideally, shoe-to-surface characteristics should mediate between players' performance requirements and respective load management. However, such dual tasks are not trivial by nature and may even be considered conflicting. It has been shown that foot translation corresponds negatively to ankle loading (Müller *et al.*, 2010a). This is

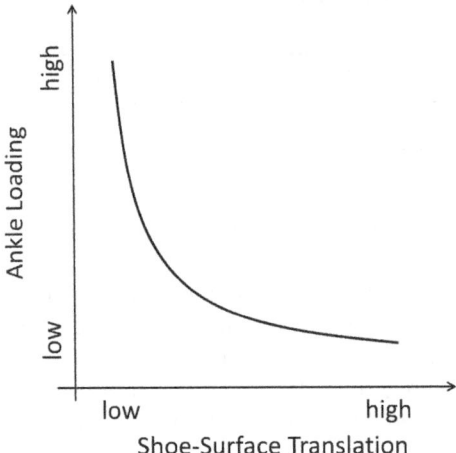

Figure 12.2 Shoe–surface translation and ankle loading during a soccer turning movement. Derived from Müller *et al.* (2010a)

illustrated by the reciprocal relationship between the magnitude of shoe–surface translation and the corresponding magnitude of ankle loading during soccer turning movements (Figure 12.2).

Regarding performance enhancement, minimal shoe–surface translation is desired so that players can maintain a consistently active motor control situation. When actions show large shoe-to-surface translation, players face a rather passive, sliding-like, motor control situation. In contrast, regarding player loading and injury prevention, minimal ankle loading is desired, illustrating the commonly referred to conflict between performance and injury prevention. Therefore, the functional performance threshold of shoe–surface translation needs to be determined, above which soccer cutting and turning performance meaningfully decreases. Furthermore, the functional loading threshold of ankle loading needs to be determined, above which players' injury risk meaningfully increases. These thresholds would provide the boundaries within which functional soccer traction should be located and should lead future soccer shoe and surface design and development efforts.

When players adapt their ground contact characteristics in an effort to mediate altered shoe–surface characteristics, they adjust rather distal lower extremity joint kinematics. In particular during initial foot strike, kinematics at the shoe-to-surface interface and at the ankle joint were adjusted, whereas knee joint kinematics appeared rather stable across various shoe-to-surface conditions. It was shown that lower grip between soccer shoes and surface caused players to adjust their foot strike more vertical to the ground in an effort to avoid slippage (Müller *et al.*, 2010a). It seems players aim to cope with altered environmental characteristics close to the source of alteration, potentially in order to keep more proximal anatomical structures of the lower extremity and the upper body alike.

Lower extremity loading of soccer players has been shown to differ for proactive and reactive actions. Thereby, proactive actions allow players more time for preparation of a specific movement, like a sudden acceleration, or a rapid cut or turn when performing feints. In contrast, reactive actions force players to act under time constraints, when the suddenly altered movement path of an opponent or a deflected ball needs to be followed, thus requiring rapid changes of previous movement directions. Advanced anticipation of game scenarios allows minimizing but not ignoring the time constraints of reactive actions. Naturally, reactive soccer specific movements are characterized by more rapid and intense braking components as proactive ones. Alongside, lower extremity loading increases during reactive compared to proactive soccer specific actions. However, regarding the functionality of soccer shoes it was observed that active and reactive soccer specific actions benefit from similar shoe–surface characteristics (Müller *et al.*, 2013).

4. Mechanical availability and biomechanical utilization of traction

During traction analyses of shoe and surface combinations, there needs to be a distinction between mechanical availability and biomechanical utilization of traction (Sterzing *et al.*, 2008; Müller *et al.*, 2010b; Luo *et al.*, 2011). In both evaluation scenarios measured values are sensitive to alteration of shoe–surface combinations. Mechanical availability of traction quantifies relative motion characteristics of shoes and surfaces where commonly the player is assumed a non-adaptive entity. It solely describes the interaction of shoes and surfaces obtained by purely mechanical measurement setups. In contrast to athlete testing, such evaluation methods commonly follow a programmed load sequence not capable of simulating physiological athlete adaptation. However, athlete adaptations are complex and highly important during assessment of functional athletic movements. These adaptations are respected and quantified during biomechanical measurements. In this sense, biomechanical utilization of traction refers to the truly functional traction a player generates during soccer specific actions. Thus, measured biomechanical values indicate the actual performance level a player achieves based on the mechanical traction properties available. Importantly, the relationships between mechanical availability of traction and its corresponding biomechanical utilization are not linear. Thus, they do not state that higher mechanical availability of traction always translates into higher biomechanical utilization. Schematically, the relationship shows a common increase of mechanical availability and biomechanical utilization of traction for shoe–surface scenarios offering rather low grip to players. The relationship then plateaus and can be assumed reversed for shoe–surface scenarios offering rather high grip to players. This indicates that above certain functional thresholds higher mechanical traction does not lead to enhanced biomechanical utilization, subsequently also causing a decrease of athletic performance (Figure 12.3, left).

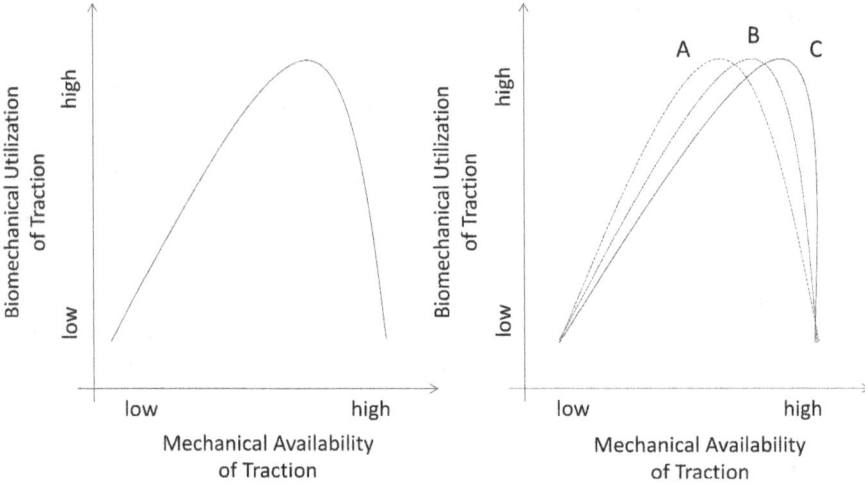

Figure 12.3 Schematic relationship between mechanical availability of traction and its biomechanical utilization (left), relationships adjusted to player subgroups (A, B, C) of different bodily characteristics (right)

It is suggested to reconsider the assumption that the same shoe–surface combination is beneficial to all players, while disregarding their individual bodily characteristics. Instead, it is recommended increased attention be paid to the players' specific bodily predisposition, triggering certain fine tuning requirements in order to design their individually optimal outsole configuration. With reference to the general relationship between mechanical availability and biomechanical utilization of traction, it appears that the magnitude of biomechanical utilization of traction is dependent on the individual player characteristics. The concept suggests that physiologically stronger and better coordinated players are better able to use higher degrees of mechanical traction. From this perspective, different players would benefit from differently tailored magnitudes of mechanical traction in order to reach their individual maximum of utilized biomechanical traction (Figure 12.3, right). Therefore, the mainly relevant bodily characteristics of players are their anthropometrics, their body disposition, their motor skills, and their training status. In consequence certain player subgroups, and in even more specialized scenarios the individual player, benefit from individualized shoe–surface traction conditions, enabling them to unfold their maximum biomechanical and athletic performance. In this respect, it is important to eventually direct all design and development efforts of soccer traction towards a maximization of biomechanically utilized traction, whereas mechanical available traction is surely regarded as a valuable and supportive tool.

Players can adapt the execution of their movements to the traction circumstances voluntarily only when they are aware of them. Below the threshold of cognitive awareness, movements can only be modified by automated spinal reflex responses, which do not involve higher cerebral levels. When aware of altered traction circumstances, soccer players have been shown to adapt and modify their movements. When the grip of shoe–surface combinations is perceived low, indicating only little mechanical availability of traction, players reduced the explosiveness of their movements and negotiated these inadequate traction circumstances by a more vertically aligned foot strike (Müller *et al.*, 2009b; Müller *et al.*, 2010a). Such coping strategies were observed when soccer shoes had their studs fully removed but are assumed to generally apply when shoe–surface combinations offer inadequate low grip to players. In contrast, when the grip of shoe–surface combinations is inadequately high, players also reduced the explosiveness of their movements, in an effort to apply a more cautious, protective foot strike (Sterzing *et al.*, 2008; Sterzing *et al.*, 2010). This coping strategy was observed when soft ground stud designs were inadequately used on artificial grass turf. The reduction of movement explosiveness is an important means for players to cope with inadequate traction circumstances, which is often accompanied by altered, more secure alignment of lower extremity kinematics. However, such voluntarily driven movement behaviour is dependent on actual player awareness of inadequate traction circumstances. Thus, it marks an inherently triggered injury protection mechanism, which so far was observed in investigations based on standardized laboratory settings. It is challenging to verify such mechanisms during actual games or during practice, as it is likely that under competition stress players' attention towards inadequate traction circumstances is reduced, in turn also reducing their inherent protective capacity. This emphasizes that the right selection of soccer shoes for a given surface is important in order to keep player loading below functional boundaries.

The underlying conceptual thoughts regarding the relationship of mechanically available and biomechanically utilized traction suggest that performance enhancement and injury prevention are not necessarily conflicting entities when looking at soccer traction. It rather seems that superior soccer traction conditions are within reach, which would serve the purpose of performance increase and load management simultaneously. When players cannot negotiate excessively high mechanically available traction thus being unable to fully transfer it into biomechanically utilized traction, and when players' bodily characteristics determine traction conditions to be adequate or inadequate, it is likely that optimal traction conditions rather resemble a function of matching shoe outsole characteristics to given surface characteristics under more precise consideration of individual players' bodily dispositions. This concept should lead the design and development of soccer shoe outsole construction to a higher level of functional individualization. Currently upcoming 3D printing technologies appear well suited to foster higher degrees of functional soccer shoe individualization. Therefore, design and development processes may be initiated by adapting soccer shoe stud configurations to player subgroups regarding their sex, age, and training status, taking into account their respective bodily predisposition as described in previous works (Morag & Johnson, 2001; Sterzing & Althoff, 2010; Althoff & Hennig, 2014).

5. Injury prevention by shoes, surfaces and training

The role of soccer shoes and the degree of traction they exhibit in interaction with the surface is frequently discussed regarding injury occurrence in soccer. However, the soccer shoe, in contrast to surfaces, as causal factor of soccer injuries has not yet been the topic of systematic research. So far, soccer injuries have been categorized regarding type and frequency (Roos *et al.*, 1995; Bjordals *et al.*, 1997; Agel *et al.*, 2005; Dick *et al.*, 2007; Waldén *et al.*, 2011), sex and age effects (Arendt & Dick, 1995; Aoki *et al.*, 2010), occurrence during match and practice (Fuller *et al.*, 2007a; Fuller *et al.*, 2007b), and also for surface effects (Engebretsen & Kase, 1987; Arnason *et al.*, 1996; Ekstrand *et al.*, 2006; Fuller *et al.*, 2007b; Steffen *et al.*, 2007). Evidence based knowledge on the potential influence of soccer shoes would provide important insight into the underlying mechanisms of soccer injury occurrence, guiding future enhancement of their injury protection capacity. At this stage it is mandatory to point out that soccer shoes should not be referred to as the sole cause of injuries. Instead they clearly need to be seen as preventive measures to protect players from injuries (Lake, 2000). Not wearing soccer footwear during competitive play can surely be assumed to largely increase injury rates, next to decreasing playing performance. Thus, it needs to be criticized when soccer shoes receive unreflected negative and sometimes even causative connotation in the discussion of injury occurrence. Traction characteristics are addressed as overly aggressive, locking the foot to the surface excessively tightly, and blocking physiologically needed translational and rotational movement amplitudes of the lower extremity. However, evidence based soccer injury surveys have not yet investigated the soccer shoe as confounding factor and thus reliable data on soccer shoe influence on injury occurrence is lacking.

Regardless of pending efforts to analyse the role of soccer shoes during occurrence of injuries, it is worthwhile to improve soccer injury prevention efforts. When injuries occur, the inner resistance of players' bodily structures is lower than externally induced forces, originating at the shoe-to-ground interface. Thereby, the exact resistance threshold of inner anatomical structures is an individual marker of player subgroups or even of each individual player. In this sense the capacity of the players' neuro-muscular and skeletal system defines and limits the anatomical and coordinative-motor resources available to prevent injuries. These are determined by the player's anthropometrics and body composition, his training status and motor skills, and are considered as given once play is started. In this sense, injury prevention training targets the enhancement of the inner resistance threshold that players' anatomical, physiological, and coordinative-motor structures provide. Therefore, anthropometric characteristics (with exception of youth players' natural growth development) do not exhibit potential of beneficial alteration. In contrast, body composition exhibits some potential, the training status good potential, and coordinative-motor capacity high potential of beneficial alteration to subsequently benefit injury prevention of players. Related efforts have been successfully implemented to strengthen players' physical predisposition for avoidance of injuries by applying specific warm-up exercises, for instance the FIFA 11+ programme (www. fifa.com), is foremost aimed at recreational and youth players. The effectiveness of such injury prevention programmes is based on improved specific muscular strength and coordination skills of players. In consequence, their inner anatomical resistance threshold is exceeded and fewer soccer specific actions are potentially hazardous to players.

In order to deal with inadequate traction soccer players also use certain inherent sensory-motor warning mechanisms. When traction circumstances are overly low, players commonly alter their foot strike angular characteristics to achieve a more vertical foot strike and to reduce the risk of slippage (Müller *et al.*, 2010a). When traction circumstances are overly high players commonly reduce their movement explosiveness to increase injury protection (Sterzing *et al.*, 2010). In both scenarios, agility running performance is decreased, corresponding to the altered movement characteristics. However, these observations were obtained during laboratory experiments when players were aware of the shoe-to-surface traction characteristics to be experienced. They also focussed solely on the required cutting or turning movement task, resembling ideal operating circumstances of the inherent sensory-motor warning mechanism. The mechanism is based on the physiological sensation and cognitive perception of external stimuli (Handwerker, 2000), which are received and processed, subsequently leading to altered motor movement responses. Therefore, potentially hazardous movements are redirected and fine-tuned according to environmental demands in an effort of injury prevention. During game and practice conditions the induced physiological and psychological stress is increased compared to laboratory settings. Anticipation of non-physiological actions may be reduced or ignored when technical and tactical demands of the game widely occupy a player's sensory-motor system. In such scenarios injuries may occur despite players having undergone high level injury prevention programmes already featuring enhanced anatomical and physiological as well as increased neuro-muscular motor capacities.

Regarding the selection of the right combinations of soccer shoes and surfaces, it was pointed out that switching between surface types is likely to contribute to injury occurrence, especially when time intervals for adequate player adjustment are relatively short (Ekstrand *et al.*, 2006; Sterzing, 2013). Such scenarios are present when soccer clubs within one league play on considerably different surface types, for instance natural grass and artificial grass surfaces. In this respect, also soccer tournaments, for instance World Cups, being played at different venues featuring different surface conditions, should be reconsidered, as they require surface switches on highest level of play within only a few days. When such scenarios cannot be avoided, thorough adjustment to the surface condition to play on next is highly important. Besides adequate selection and proper adjustment to shoe–surface conditions, innovative training initiatives are recommended to increase performance and injury prevention of players (Sterzing, 2016). The proposed training concept aims to go beyond currently applied training programmes, focussing on enhancing the anatomical and physiological disposition of players for better mediation of body loading. Most of the currently applied training programmes foremost incorporate expected ground contact scenarios, which players are able to anticipate. However, when players are fully aware of the nature of loading scenarios to be faced, they are reasonably well capable of mediating them and protecting themselves. Thus, recommended innovative training concepts should rather aim to apply unexpected ground contact scenarios in order to build up new motor programmes, enhancing the fundamental motor capacity of players (Sterzing, 2016). In this sense, the goal of such recommended training regimens is the increase of internally available movement responses of players. The introduced training approach follows insight into motor control theories, indicating that a larger plurality of motor performance skills also increases general human locomotion quality (Latash, 2012). Therefore, the proposed training concept can be executed by applying constantly varying surface conditions, providing unexpected ground contact circumstances to players (Sterzing *et al.*, 2014a; Sterzing *et al.*, 2014b), or by applying constantly varying shoe conditions, providing unexpected ground contact circumstances to players (Apps *et al.*, 2015a; Apps *et al.*, 2015b).

In general, injury prevention efforts need to be approached from multiple angles. Players need to be in adequate anatomical and physiological condition, the shoe-to-surface interaction needs to aim for minimal necessary levels of available mechanical traction, while enabling players to utilize biomechanical traction without compromising their playing performance. Additionally, training programmes addressing the enhancement of internally available motor locomotion programmes provide a beneficial tool to further improve injury prevention of soccer players.

6. Conclusions

Regarding functional soccer traction, the players' bodily characteristics, the game specific action, the surface type played on, and the soccer shoe outsole configuration must be considered during respective evaluation and optimization. In a competitive sport like soccer, performance and injury prevention aspects are mandatory and need to be included in functional analyses of shoe–surface alteration. However, as research indicated that overly high degrees of mechanically available traction appear dysfunctional and do not lead to higher performance, the general thinking that performance and injury prevention are conflicting entities does not hold true in all scenarios. Instead, it is proposed that the highest degree of mechanically available traction an individual player can biomechanically utilize also corresponds to the level of his sustainable physiological loading. Next to inadequate selection of shoe–surface conditions for player subgroups or individual players, it needs to be considered whether game and practice induced stress creates instances of decreased cognitive and sensory awareness, subsequently leading to certain traumatic injuries.

It is suggested that soccer traction categories for player subgroups of sex (Sterzing & Althoff, 2010) and age (Morag & Johnson, 2001), as well as for individual players (Sterzing, 2016) are established (Figure 12.4). These soccer traction

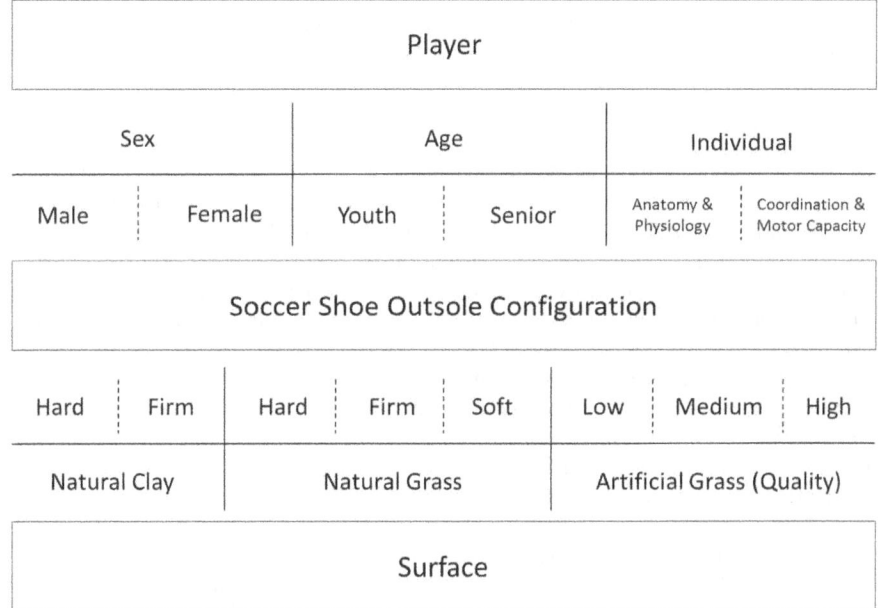

Figure 12.4 Factors to consider for soccer shoe outsole constructions

categories should offer specifically defined amounts of mechanically available traction, which should match the physiological and coordinative capacities of players. Therefore, a category offering lower mechanical traction would be recommended in unclear scenarios. As the soccer shoe mediates between the player and the surface, it probably offers largest and most flexible opportunities to influence soccer traction characteristics, which need to be reflected in future design and development efforts. It is noteworthy that due to the number of surface types that soccer, especially on sub elite level, is played on, manufacturers increasingly aim to also create soccer shoe outsoles suited for playing on different surface types. These are referred to as multi-ground soccer shoes and challenge the traditional approach of providing traction functionality to players for one specific surface type only.

7. Acknowledgements

The content of this chapter was retrieved from scientific literature and the author's personal athletic footwear research, carried out at the following academic institutions and athletic footwear companies: University of Duisburg-Essen (Germany) in collaboration with Nike Inc. (Beaverton, USA), Chemnitz University of Technology (Germany) in collaboration with Puma SE (Herzogenaurauch, Germany), Li Ning Co Ltd (Beijing, China), and Xtep Co Ltd (Xiamen, China).

References

Agel, J., Arendt, E. A. & Bershadsky, B. (2005). Anterior cruciate ligament injury in national collegiate athletic association basketball and soccer: A 13-year review. *American Journal of Sports Medicine*, 33(4), 524–530.

Althoff, K. & Hennig, E. M. (2014). Criteria for gender-specific soccer shoe development. *Footwear Science*, 6(2), 89–96.

Aoki, H., Kohno, T., Fujiya, H., Kato, H., Yatabe, K., Morikawa, T. & Seki, J. (2010). Incidence of injury among adolescent soccer players: A comparative study of artificial and natural grass turfs. *Clinical Journal of Sport Medicine*, 20(1), 1–7.

Apps, C., Sterzing, T., O'Brien, T., Cheung, J. & Lake, M. (2015a). Lower limb biomechanics during the forward lunge: The influence of a shoe with irregular deformations of the midsole. In *Proceedings of the 25th Congress of the International Society of Biomechanics*, Glasgow, UK.

Apps, C., Sterzing, T., O'Brien, T., Cheung, J. & Lake, M. (2015b). Irregular shoe midsole deformations simulate the movement variability of running on uneven surface. *Footwear Science*, 7(Suppl 1), 12.

Arendt, E. & Dick, R. (1995). Knee injury patterns among men and women in collegiate basketball and soccer. *American Orthopaedic Society for Sports Medicine*, 23(6), 694–701.

Arnason, A., Gudmundsson, A., Dahl, H. A. & Jóhannsson, E. (1996). Soccer injuries in Iceland. *Scandinavian Journal of Medicine & Science in Sports*, 6(1), 40–45.

Bjordals, J. M., Arnøy, F., Hannestad, B. & Strand, T. (1997). Epidemiology of anterior cruciate ligament injuries in soccer. *American Journal of Sports Medicine*, 25(3), 341–345.

De Clercq, D., Debuyck, G., Gerlo, J., Rambour, S., Segers, V. & Van Caekenberghe, I. (2014). Cutting performance wearing different studded soccer shoes on dry and wet artificial turf. *Footwear Science*, 6(2), 81–87.

Dick, R., Putukian, M., Agel, J., Evans, T. A. & Marshall, S. W. (2007). Descriptive epidemiology of collegiate women's soccer injuries: National collegiate athletic association injury surveillance system, 1988–1989 through 2002- 2003. *Journal of Athletic Training*, 42(2), 278–285.

Ekstrand, J., Timpka, T. & Hägglund, M. (2006). Risk of injury in elite football played on artificial turf versus natural grass: Prospective two-cohort study. *British Journal of Sports Medicine*, 40(12), 975–980.

Engebretsen, L. & Kase, T. (1987). Soccer injuries and artificial turf. *Tidsskr Nor Laegeforen*, 107, 2215–2217.

Fédération Internationale de Football Association (2015). Laws of the game 2015–2016, Zurich, Switzerland.

Fuller, C. W., Dick, R. W., Corlette, J. & Schmalz, R. (2007a). Comparison of the incidence, nature and cause of injuries sustained on grass and new generation artificial turf by male and female football players. Part 1: Match injuries. *British Journal of Sports Medicine*, 41(Suppl, 1), 20–26.

Fuller, C. W., Dick, R. W., Corlette, J. & Schmalz, R. (2007b). Comparison of the incidence, nature and cause of injuries sustained on grass and new generation artificial turf by male and female football players. Part 2: Training injuries. *British Journal of Sports Medicine*, 41(Suppl I), 27–32.

Handwerker, H. (2000). Allgemeine Sinnesphysiologie. In *Physiologie des Menschen* (pp. 195–235), Springer, Berlin Heidelberg, Germany.

Hennig, E. M. & Sterzing, T. (2010). The influence of soccer shoe design on playing performance – a series of biomechanical studies. *Footwear Science*, 2(1), 3–11.

Krahenbuhl, G. (1974). Speed of movement with varying footwear conditions on synthetic turf and natural grass. *Research Quarterly*, 45(1), 28–33.

Lake, M. (2000). Determining the protective function of sports footwear. *Ergonomics*, 43(10), 1610–1621.

Latash, M. L. (2012). The bliss of motor abundance. *Experimental Brain Research*, 217(1), 1–5.

Luo, G. & Stefanyshyn, D. J. (2011). Identification of critical traction values for maximum athletic performance. *Footwear Science*, 3(3), 127–138.

Morag, E. & Johnson, D. (2001). Traction requirements of young soccer players, In *Proceedings of the 5th Symposium of Footwear Biomechanics*, 62–63.

Müller, C., Sterzing, T., Kunz, A. & Milani, T.L. (2009a). Soccer game characteristics and players' perception on artificial turf and natural grass. In *Proceedings of the 22th Congress of the International Society of Biomechanics*, Available from: https://isbweb. org/images/conf/2009/data/pdf/99.pdf.

Müller, C., Sterzing, T. & Milani, T.L. (2009b). Stud length and stud geometry of soccer boots influence running performance on third generation artificial turf. In A. J. Harrison, R. Anderson & I. Kenny (Eds.) *Proceedings of the 27th Conference of the International Society of Biomechanics in Sports* (pp. 811–814), Limerick: University of Limerick.

Müller, C., Sterzing, T., Lake, M. & Milani, T. L. (2010a). Different stud configurations cause movement adaptations during a soccer turning movement. *Footwear Science*, 2(1), 21–28.

Müller, C., Sterzing, T., Lange, J. & Milani, T. L. (2010b). Comprehensive evaluation of player-surface interaction on artificial soccer turf. *Sports Biomechanics*, 9(3), 193–205.

Müller, C., Sterzing, T. & Milani, T. L. (2013). Unanticipated compared to preplanned turning movements increase lower extremity loads in soccer. In H. Nunome, B. Drust & B. Dawson (Eds.) *Science and Football VII* (pp. 21–26), Routledge, London New York.

Roos, H., Ornell, M., Gärdsell, P., Lohmander, L. S. & Lindstrand, L. S. (1995). Soccer after anterior cruciate ligament injury – an incompatible combination? A national survey of incidence and risk factors and a 7-year follow up of 310 players. *Acta Orthopaedica Scandinavica*, 66(2), 107–112.

Schrier, N., Wannop, J., Lewinson, R., Worobets, J. & Stefanyshyn, D. (2014). Shoe-surface interaction affects performance of soccer-related movements. *Footwear Science*, 6(2), 69–80.

Smith, G., Lake, M., Sterzing, T. & Milani T.L. (2016). Influence of sprint spike bending stiffness on sprinting performance and metatarsophalangeal joint function. *Footwear Science*, 8(1), 109–118.

Stefanyshyn, D. J. & Fusco, C. (2004). Increased shoe bending stiffness increases sprint performance. *Sports Biomechanics*, 3, 55–66.

Steffen, K., Andersen, T. E. & Bahr, R. (2007). Risk of injury on artificial turf and natural grass in young female football players. *British Journal of Sports Medicine*, 41(Suppl I), 33–37.

Sterzing, T. (2013). Injury occurrence and footwear performance on artificial soccer turf. In H. Nunome, B. Drust & B. Dawson (Eds.), *Science and Football VII* (pp. 9–14), Routledge: London New York.

Sterzing, T. (2015). Athletic footwear research: Effects of shoe construction and relationships of evaluation parameters, *Habilitation*, University of Duisburg-Essen, Essen, Germany.

Sterzing, T. (2016). Der Fußballschuh der Zukunft: Wechselwirkung von Schuh, Untergrund und Training, 3. *DFB-Wissenschaftskongress – Fußball im Spannungsfeld zwischen Meisterlehre und Evidenz, Frankfurt am Main*, Deutschland, 66–69.

Sterzing, T. & Althoff, K. (2010). Begründung eines Frauenfußballschuhs, *Orthopädieschuhtechnik*, 6, 22–27.

Sterzing, T., Apps, C., Ding, R. & Cheung, J. (2014a). Walking on an unpredictable irregular surface changes lower limb biomechanics and subjective perception compared to walking on a regular surface. *Journal of Foot and Ankle Research*, 7(Suppl 1), A64.

Sterzing, T., Apps, C., Ding, R. & Cheung, J. (2014b). Running on an unpredictable irregular surface changes lower limb biomechanics and subjective perception compared to running on a regular surface. *Journal of Foot and Ankle Research*, 7(Suppl 1), A63.

Sterzing, T. & Hennig, E. M. (2008). The influence of soccer shoes on kicking velocity in full instep soccer kicks. *Exercise and Sport Sciences Reviews*, 36(2), 91–97.

Sterzing, T., Lam, W.K. & Cheung, J. (2012). Athletic footwear research by industry and academia. In R. S. Goonetlleke (Ed.), *The Science of Footwear* (pp. 605–622), Boca Raton, FL: CRC Press Taylor and Francis.

Sterzing, T., Müller, C., Hennig, E. M. & Milani, T.L. (2009). Actual and perceived running performance in soccer shoes: A series of eight studies. *Footwear Science*, 1(1), 5–17.

Sterzing, T., Müller, C. & Milani, T.L. (2010). Traction on artificial turf: Development of a soccer shoe outsole. *Footwear Science*, 2(1), 37–49.

Sterzing, T., Müller, C., Schwanitz, S., Odenwald, S. & Milani, T. L. (2008). Discrepancies between mechanical and biomechanical measurements of soccer shoe traction on artificial turf. In Y-H. Kwon, J. Shim, J. K. Shim & I-S. Shin (Eds.) *Proceedings of the 26th Symposium of the International Society of Biomechanics in Sports* (pp. 339–342), Seoul: Korean Society of Biomechanics & Seoul National University.

Toon, D., Williams, B., Hopkinson, N. & Caine, M. (2009). A comparison of barefoot and sprint spike conditions in sprinting, *Journal of Sports Engineering and Technology, Proceedings of the Institute of Mechanical Engineers*, Part P, 223, 77–87.

Waldén, M., Hägglund, M., Werner, J. & Ekstrand, J. (2011). The epidemiology of anterior cruciate ligament injury in football (soccer): a review of the literature from a gender-related perspective, Knee Surgery, Sports Traumatology, *Arthroscopy*, 19(1), 3–10.

13 Boot–turf interaction during a 180° cutting movement on artificial turf when wet and dry

Dirk De Clercq, Gijs Debuyck, Alison Sheets-Singer, Joeri Gerlo, Stijn Rambour, Veerle Segers and Ine Van Caekenberghe

1. Introduction

The ability to perform fast cutting maneuvers is essential for success in soccer. These cutting maneuvers are characterized by substantial changes in speed and/or direction, thus requiring large horizontal forces and impulses to be exerted by the feet on the surface. Sufficient traction created by the interaction of the field and shoe is necessary to ensure that a player can successfully perform the movement with minimal slipping (Lake, 2000). Traction, along with stability and comfort, has also been among the most consistently requested attributes for soccer shoes by male players (Hennig & Sterzing, 2010).

Traction generated between the shoe and surface can be quantified both mechanically and biomechanically (Sterzing *et al.*, 2008, Potthast *et al.*, 2010, Luo & Stefanyshyn, 2011, McGhie & Ettema, 2013). The mechanically available traction is typically quantified using a machine to measure the resistance of the shoe (or portion of the shoe) to movement relative to the testing surface. Utilized traction is typically measured during biomechanical tests in which athletes perform maximal effort/speed movements which require the generation of large horizontal forces for successful completion. Numerous movements including accelerations, decelerations, cutting, and turning movements, with varying degrees of game-realism have been performed to biomechanically test utilized traction. Some experiments perform the movements in isolation, and others link them together to create "Functional Traction Courses" (Hennig & Sterzing, 2010).

Increasing the mechanically available traction between the shoe and surface has been shown to improve change of direction performances (Müller *et al.*, 2010) until a threshold minimum mechanical traction value has been exceeded. Once the minimum traction has been achieved, additional traction was not shown to continue to improve performance, and the biomechanically utilized traction was lower than the mechanically available traction that the footwear and surface could provide (Luo & Stefanyshyn, 2011).

Many factors can influence the mechanical traction of the soccer shoe–turf surface interface including the playing surface, footwear outsole attributes, and the environment, and individual player biomechanics can influence the utilized traction (Severn *et al.*, 2008). Artificial turf (AT) surfaces are widely used in soccer (e.g. through the FIFA Quality Concept for Football Turf (2009)). There are 25 AT manufacturers that provide turf to FIFA, and 2,508 turf installations around the world that achieved at least a quality rating (comparable to FIFA * rating) when this chapter was written (FIFA Quality Programme 2016). These third-generation AT surfaces are differentiated from previous generations by the inclusion of a loose infill between the synthetic fibers. An example infill is a sand base layer with a rubber particle layer on top, although other AT surfaces may vary the materials, ratio of materials, and/or layering of infill materials. These third-generation turf surfaces also utilize different fiber materials, shapes, lengths, and densities, all of which may affect the mechanical traction between the soccer shoe and turf surface.

Although AT surface moisture level and outsole stud configuration of the soccer shoe have been shown to affect mechanical traction, few studies have performed biomechanical experiments to investigate the effects on performance. Soccer players commonly report experiencing less grip in wet AT conditions, such as those occurring after a rain or after an AT field is watered, and previously reported mechanical traction values for wet and dry surfaces support this feedback (Heidt *et al.*, 1996). This could be due to a shoe–turf boundary lubrication effect created by the wet rubber infill particles. Players also commonly choose shoes with an outsole containing many, relatively short studs for playing on dry AT. But on wet AT, players may choose longer and/or fewer studded outsole designs in order to have greater penetration of the infill and interaction with the fibers. However, in a pilot study with five soccer players, Sterzing and colleagues (2009) found that on wet AT a soft ground stud configuration (i.e. with 6 long studs and suited for playing on soft natural grass) did not provide superior traction related functionality compared to hard and firm ground stud configurations with more, moderate length studs. Therefore, it is relevant to investigate moisture induced differences in traction related performance by performing a study with subjects wearing shoes with outsole configurations that players commonly choose for playing on AT. It has also been shown that experienced soccer players can perceive shoe–surface induced differences in traction (e.g. Sterzing, 2009), but we are unaware of experiments evaluating whether players can perceive traction differences between dry and wet artificial turf conditions.

Using experienced soccer players, the aim of this study was to quantify the influence of moisture (dry and wet AT) and three commonly chosen shoe stud configurations (Turf Field (TF), Artificial Grass (AG) and Firm Ground (FG)) on maximal effort cutting performance, utilized traction during the cut, and players' perception.

2. Methods

Twelve experienced soccer players with active player status in a Belgian first division club (age 16 ± 1 years; stature 176.3 ± 8.8 cm; body mass 67.3 ± 8.1 kg; shoe size US 9.5 ± 1.5) participated in this study, which was approved by the

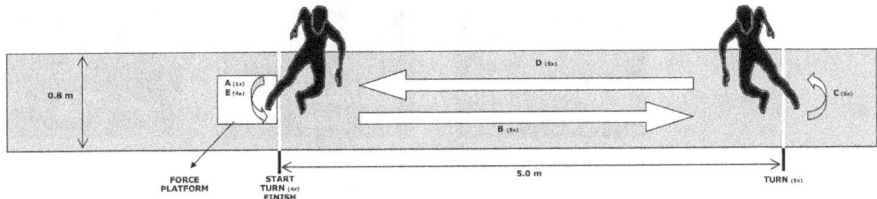

Figure 13.1 Experimental set-up of the shuttle run test (SHR). Players start (A) from a standing position behind the start line, (B) run 5m to the turning line, (C) place one foot behind this line and turn 180°, (D) run back to the starting line (E) place one foot behind this line and 180° and repeat (B) to (E) four more times. During this last repetition at (E) the players continue to run over the line. Athletes are asked to execute the drill as fast as possible on different two turf wetness conditions, and in three different shoes conditions (6 combinations)

ethical committee of Ghent University Hospital. In this experiment, players performed two maximal effort 10 × 5 m shuttle runs in each condition (SHR) (see Figure 13.1) to measure footwear/turf traction related changes in cutting performance. This test was chosen because successfully completing the nine high speed, 180° turns requires large shear forces to be created between the cleats and turf (Shorten *et al.*, 2003), and because it is a reliable measure for fast acceleration, deceleration and turning agility since it is frequently used in soccer (Council of Europe, 1988).

During this experiment, dependent variables measured were player performance, utilized traction, and player perception. Performance was measured using time to complete the SHR drill, traction was quantified from the forces generated during the 180° turns on the force platform, and perception was measured using a visual analog scale questionnaire. Two different turf wetness conditions (dry or wet AT) x 3 different outsole designs (shoe outsole, Figure 13.2) were tested using a block randomized design. All 6 combinations of turf wetness and footwear conditions were performed in the same day, and the beginning surface wetness condition was randomized and balanced between subjects (i.e. half of the subjects started with the dry condition and vice versa).

Tests were performed on wet (1.5 L m^{-2}) and DRY Desso Challenge Pro^2 artificial grass (6 cm fibre length) with SBR-infill (4 cm). The correct amount of water for the given surface area was sprayed on the entire length of the artificial turf at the beginning of the experiment and qualitatively maintained throughout the experiment i.e. entirely dry or entirely wet. Mechanical tests for impact, rotational resistance and linear friction fulfilled the FIFA recommended 2 star requirements (FIFA, 2009), but did not differentiate between dry and wet AT conditions. It is possible that the FIFA mechanical tests did not differentiate the wetness conditions because the testing procedure prescribes more water immersion for the wet testing condition.

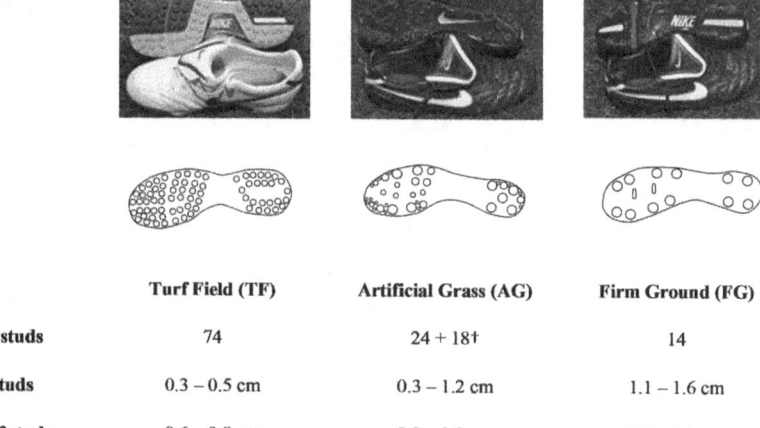

	Turf Field (TF)	Artificial Grass (AG)	Firm Ground (FG)
Number of studs	74	24 + 18†	14
Length of studs	0.3 – 0.5 cm	0.3 – 1.2 cm	1.1 – 1.6 cm
Diameter of studs	0.6 – 0.8 cm	0.3 – 1.2 cm	0.8 – 1.1 cm

Figure 13.2 Three Nike Tiempo Mystic III models with different artificial turf outsole stud characteristics were tested.† In the AG model, these 18 studs are very small compared to the other 24

Players wore the Nike Tiempo Mystic III model with three different types of shoe stud characteristics ranked from small to average stud length: Turf Field (TF), Artificial Ground (AG) and Firm Ground (FG) sole. It is reasonable to assume that players/consumers might choose to wear any of these three outsole designs on AT, therefore, the supposition was made that they present a realistic range of available traction.

Performance was quantified by measuring the time to complete each of the maximal effort 10 × 5 m shuttle runs. The beginning of the shuttle run time was defined as the moment when the subject last contacted the force platform (Figure 13.1, portion A), and the end of the shuttle run time was defined as the first contact on the force platform at the finish (Figure 13.1, portion E). In the few cases when no contact was made on the force platform at the finish, the time when the center of mass of the hip segment crossed the finish line was measured, using motion capture data (Qualisys Pro Reflex, 200 Hz). This alternate measurement procedure yielded a minimal accuracy of 0.01 s. If subjects slipped and/or lost balance during the test, it was noted and the results were not included in the analysis of the performance and the subjects were asked to complete the trial again. Players performed two shuttle runs in each outsole design–moisture combination. The intra-class correlation coefficient (ICC) for these two trials was higher than 0.87, indicating a high intra-subject repeatability. As such, the average performance for these two trials was used in the comparisons.

Ground reaction forces generated by the outside foot (Shorten *et al.*, 2003) during the 180° cuts were used to calculate the utilized traction and were measured at 1000 Hz with a 1.0 × 0.4 m AMTI (Watertown, MA, USA) force plate.

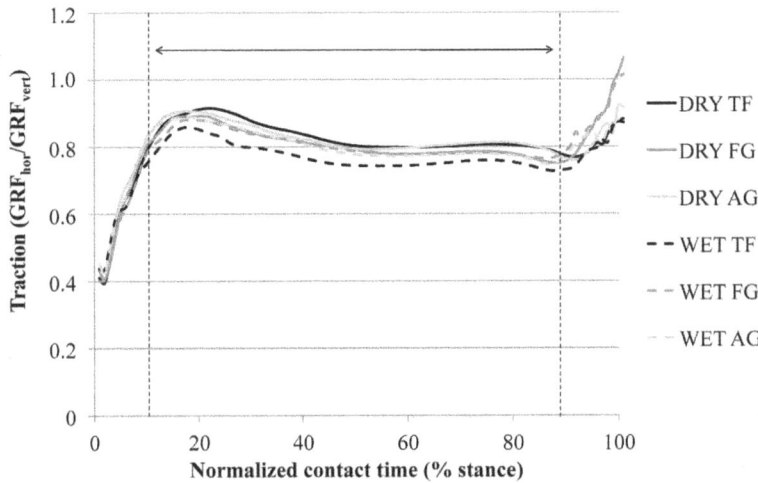

Figure 13.3 The average utilized traction produced throughout the 180° cut by the outward pivot foot, was calculated as ratio of the horizontal to the vertical component of the ground reaction force. Six conditions were measured: three shoe types Turf Field (TF), Artificial Grass (AG), Firm Ground (FG), each combined with two wetness conditions wet and dry AT. The grey area indicates the interval during which mean traction was calculated

The AT was firmly fastened on top of the plate and isolated from the surrounding AT surface. After low pass digital filtering at 50 Hz, the utilized traction was quantified using the time dependent traction ratio (Shorten *et al.*, 2003), dividing the horizontal by the vertical component of the ground reaction force as follows.

$$\text{Required traction } \tau(t) = \frac{\text{GRF total}(t)}{\text{GRF vert}(t)}$$

The traction ratio shows large variability at initial contact and at the end of the foot contact phase due to the small vertical ground reaction forces. Therefore, the average traction value was calculated starting at 10% of foot contact time and ending when the vertical component of the ground reaction force dropped under half body weight (Figure 13.3). Two SHRs were executed per condition, and five foot contacts out of the potential eight collected during the two trials were studied to test the repeatability of movement with intra-class correlation coefficients (ICC). ICC's of the average traction ratio for these five foot contacts was 0.935, meaning there was a high intra-subject repeatability between trials, which allowed averaging per subject per condition.

A specific questionnaire (visual analog scale, Figure 13.4) was used to assess the subjective perception of the players. The subjective parameters investigated were foot/ankle stability, shoe–surface grip, overall shoe comfort, rotational load and the general appreciation of the shoe. Players scored each parameter by

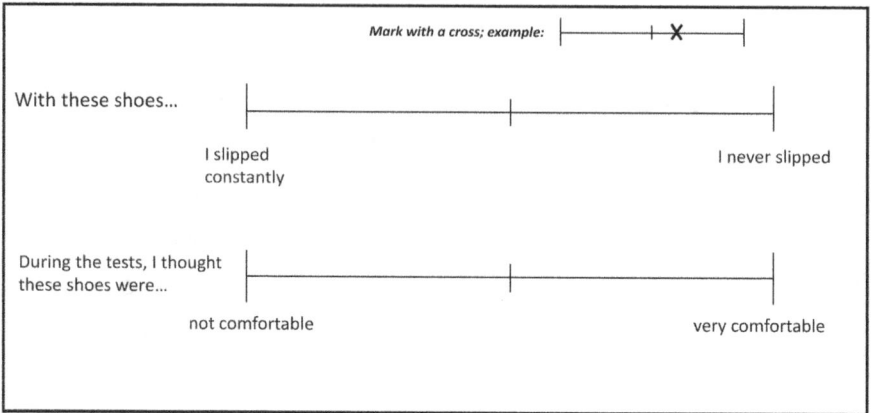

Figure 13.4 An example of a visual-analog scale that players used to provide perception feedback on different attributes of each shoe–turf combination

placing a cross on a 10 cm long line, on which the left extremity represented the worst score, and the right extremity the best score. The middle of the line was indicated by a short vertical hatch mark and represented neutrality.

Effects of the shoe and surface were statistically tested within subjects using a 3 × 2 Repeated Measures ANOVA with Bonferroni correction on the potential main shoe effect. Significance was obtained when $p < 0.05$ (SPSS 15, Chicago IL). Whenever a significant shoe × surface interaction effect was present, two additional statistical tests were conducted. Post-hoc pairwise comparisons were conducted for each shoe in the dry and wet condition (paired samples t-test), and an ANOVA with Bonferroni correction was conducted comparing the three shoes within the dry and wet conditions separately.

3. Results

3.1 Performance

There was a significant shoe x surface interaction effect on the time to complete the SHR ($p < 0.005$, Table 13.1), and a significant main effect of surface ($p < 0.005$). Pairwise comparisons indicated that only for the TF shoe, SHR time is significantly larger (0.73 s) for the wet compared to the dry condition ($p < 0.05$). Only in the wet condition, SHR time is significantly longer for the TF compared to the other two shoes (ANOVA $p < 0.05$ and pairwise $p < 0.05$). The longer SHR time with the TF shoe in the wet condition concurs with the trend for an interaction effect ($p = 0.1$) in the foot contact time; post-hoc pairwise comparisons indicated a trend ($p = 0.08$) for a longer contact time when wearing the TF shoe in the WET (0.40 ± 0.04 s) compared to the dry condition (0.36 ± 0.06 s).

Table 13.1 Performance of the 10 × 5 m shuttle run and traction ratio (mean ± SD) in each shoe – surface wetness condition tested [N = 11].

	Turf Field (TF)	Artificial Grass (AG)	Firm Ground (FG)
Performance time (s)			
DRY	14.66 ± 0.46 *	14.86 ± 0.57	14.85 ± 0.58
WET	15.39 ± 0.59 *, FG, AG	14.99 ± 0.46 TF	14.98 ± 0.53 TF
Traction ratio			
DRY	0.81 ± 0.05 *	0.81 ± 0.05 t	0.79 ± 0.05
WET	0.76 ± 0.05 *,FG,AG	0.79 ± 0.04 t, TF	0.79 ± 0.04 TF

For Performance time
RM ANOVA:
main effect shoe (p > 0.1), main effect surface (p < 0.005), shoe-surface interaction effect (p < 0.005).
INTERACTION EFFECT:
- pairwise comparison DRY-WET for each shoe: * = p < 0.05.
- ANOVA comparing shoes: DRY condition (p > 0.1) - WET condition (p < 0.05; TF, AG, FG = p < 0.05).
For Traction ratio
RM ANOVA:
main effect shoe (p > 0.1), main effect surface (p < 0.05), shoe-surface interaction effect (p = 0.06).
INTERACTION EFFECT:
- pairwise comparison DRY-WET for each shoe: * = p < 0.05, t = p < 0.1.
- ANOVA comparing shoes: DRY condition (p > 0.1) - WET condition (p < 0.05; TF, AG, FG = p < 0.05).

3.2 Utilized traction

A trend towards a shoe x surface interaction effect was observed (p = 0.06, Table 13.1), as well as a significant main effect of the surface (p < 0.05). Pairwise comparisons indicated that only for the TF shoe, utilized traction is significantly lower for the wet compared to the dry condition (p < 0.05). only in the wet condition, utilized traction is significantly lower for the TF compared to the other two shoes (ANOVA p < 0.05 and pairwise p < 0.05). Additionally, during the entire experiment 7 of the 8 slips of the outward pivot foot that resulted in a complete loss of body balance occurred in the wet x TF shoe condition.

3.3 Player perception

A significant shoe x surface interaction effect was measured for the player's perception of shoe–surface grip (p < 0.005), foot-ankle stability (p < 0.05) and general shoe appreciation (p < 0.05) (Figure 13.5 A, B, and C, respectively). Pairwise comparisons indicated that only for the TF shoe grip, stability and appreciation are lower for the wet compared to the dry condition (p < 0.05). ANOVA including pairwise comparisons indicated that only in the wet condition, a significantly worse and a trend towards a significantly worse perception was observed compared to the FG and the AG shoe (ANOVA p < 0.05, pairwise respectively p < 0.05 and p = 0.08 ± 0.01).

A significant main shoe effect (p < 0.05), combined with the pairwise comparison between all shoes, indicated that the overall shoe comfort was lower in the TF shoe compared to the other two types (p < 0.05).

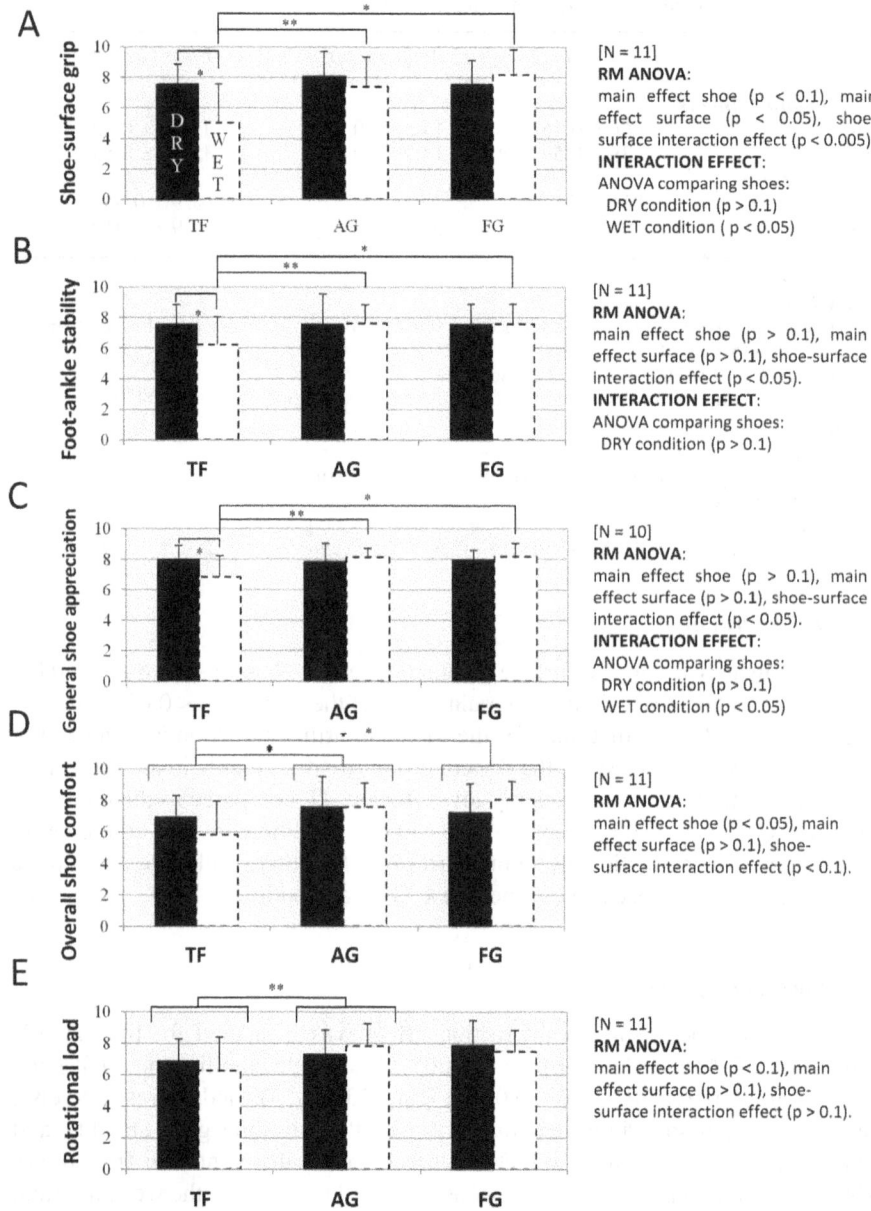

Figure 13.5 Players' perception of the three soccer shoe designs in the wet and dry condition. Error bars indicate 1 standard deviation. * $p < 0.05$, ** $p < 0.1$ (pairwise comparisons)

A trend towards a significant main shoe effect (p = 0.09), combined with a pairwise comparison between all shoes, indicated that the perception of rotational load tended to be lower for the TF shoe compared to the AG shoe (p = 0.09).

4. Discussion

Experienced soccer players executed 180° turns, during a 10 × 5 m maximal effort shuttle run, to test for differences in performance and biomechanics with changes in traction due to interaction between the soccer shoe outsole and artificial turf (AT) surface wetness. This was done for three realistic shoe outsole conditions, and a wet and dry AT surface condition. The intra-subject variability was low for the performance and traction variables, demonstrating that results obtained with this test are reliable.

The wet artificial turf condition in combination with the multi-short-studded Turf Field (TF) shoe, equipped with 74 short studs, led to a 5% worse SHR performance and players had an approximately 6% lower utilized traction. The performance reduction is comparable with (and even slightly exceeds) results of studies testing performance differences on Functional Traction Courses when players wear soccer shoes with realistic differences in stud designs (e.g. Sterzing *et al.*, 2009). The detrimental effect of moisture on utilized traction when wearing multi-studded outsole designs on third-generation AT surfaces also concurs with mechanical testing results, which demonstrated a substantial reduction in available traction of this outsole configuration on wet AT (Wannop *et al.*, 2012). Furthermore, seven of the eight real slips (out of 528 recorded trials) occurred in the TF shoe x wet AT condition. These real slips were characterized by a marked loss of balance and large performance deficit, and were not included in the statistical comparisons. These results indicate that the TF shoe outsole does not offer enough traction on wet AT to enable maximal player performance. Conversely, on wet AT compared to the dry condition, performance and utilized traction did not significantly change for the Artificial Grass (AG) and the Firm Ground (FG) outsoles, suggesting that these medium aggressive outsole designs (defined by the combination of length and number of studs) provide enough traction for maximal performance on wet AT.

In the dry condition, players performed the SHR equally well in all three shoes with similar utilized traction. This result is similar that of McGhie and Ettema (2013) who found similar utilized traction coefficients during a 90 degree cut using three cleat configurations on three different third-generation AT surfaces. These results are interesting because each model tested in this experiment had a different number of studs, with different lengths. It is possible that the mechanisms of stud and turf interaction were slightly different between the shoes, yet all provided the minimum amount of mechanical traction necessary in the dry condition.

The combination of the findings about shoe performance on wet and dry artificial turf is relevant for soccer practice. On dry or wet AT, performance and utilized traction were similar during the SHR in the AG and FG shoes, suggesting that both shoes provided a minimum amount of mechanically available traction independent of the surface wetness. While no differences were found on dry AT in this experiment,

Sterzing and colleagues (2010) found that soccer shoes with multiple relatively short stud elements provided better functional traction on dry AT.

The agreement between measured differences in performance, utilized traction, and player perception measurements reveal that experienced players could perceive performance-related traction differences. The perceived grip is indeed lower for the wet x TF condition compared to all other conditions. Interestingly, there is also a difference in perception for foot–ankle stability with the wet x TF condition. The perceived comfort of the TF shoe was lower than the other shoes independent of surface wetness condition. Finally, the general appreciation score of the TF shoe on the wet surface was worse, indicating that general appreciation indeed depended on a combination of perceived traction, stability and comfort (Hennig *et al.*, 2010).

It should be stressed that all results apply to these tested conditions, including low to medium aggressive outsole designs and a third-generation AT surface with rubber infill. It remains to be seen whether the threshold values for utilized traction on movements performed in a laboratory setting are the same as when performed in more game-realistic situations, or on different turf surfaces. It is possible that athletes would need slightly more traction in game-realistic settings, as there may be more reactive movements, movement variability, and/or less consistency in the field surface. If more traction was needed, it is possible that differences in performance between more conditions may have been found. Additionally, it is unknown whether the results from this experiment are directly applicable to other shoe and turf systems. The different components used to create the artificial turf surface (e.g. fiber density, length, shape, and/or material, infill depth and/or material, turf age and/or maintenance history, etc.) can affect mechanical traction testing results and player perception data (Potthast *et al.*, 2010).

Incorporating fast cutting maneuvers in a maximal effort shuttle run task, which soccer players are familiar with, resulted in consistent and reliable movement patterns. Consistency in the movement technique and effort level enabled a relationship to be found between differences in utilized traction and maximal performance. Nevertheless, it can be questioned whether the 180° turn performed in this experiment is frequently executed in match situations. Alternative movements that have increased traction requirements, and thus would be appropriate for evaluating the relationship between performance and traction, are cutting maneuvers with smaller directional changes (e.g. Müller *et al.*, 2010), or Functional Traction Courses that incorporate numerous different directional changes performed continuously (Sterzing *et al.*, 2009).

A methodological strength of the current study was that the whole AT surface track was transformed to the dry or wet conditions, and not only the AT surface on top of the force platform. A limitation was that movement phases other than the 180° cut, such as the accelerating and decelerating steps in the total SHR task, can also influence the global performance outcome, and utilized traction during the steps before or after the outward pivot foot contact were not measured. It is possible that players adapt in these support phases to the different turf x shoe conditions.

5. Conclusions

A within-subjects experiment was conducted to test the effect of dry and wet artificial turf conditions on performance, utilized traction, and perception for three realistic soccer shoe stud designs. The findings apply for third-generation artificial turf and the wet condition can be compared to field conditions after a light rain fall. Soccer players wearing a multi-short-studded Turf Field shoe, utilized less traction on the outward pivot foot when performing fast 180° turns on wet artificial turf compared with dry. This finding corresponded to an approximately 5% slower 10 × 5 m shuttle run performance, which incorporated nine 180° turns. Performance was not affected by wet artificial turf when players wore shoes with other common Artificial Grass or Firm Ground stud designs, both of which had fewer, longer stud lengths. Experienced players perceived these differences in performance and traction. On dry artificial turf, no differences between the three tested stud designs were measured for performance or utilized traction. From a maximal performance perspective, the results from this experiment suggest that a Turf Field shoe stud design may negatively impact performance when playing on wet AT, but did allow players enough traction on dry AT. During this experiment, shoes with stud characteristics similar to the Artificial Grass and Firm Ground shoe fulfilled the traction needs on both wet and dry artificial turf. Therefore, shoes with even more pronounced stud designs may not be necessary on dry or wet artificial turf.

Acknowledgements

This majority of the content of this article is taken from "Cutting performance wearing different studded soccer shoes on dry and wet artificial turf" (authors Dirk De Clercq, Gijs Debuyck, Joeri Gerlo, Stijn Rambour, Veerle Segers and Ine Van Caekenberghe) published online on 1 April 2014 in the *Footwear Science Journal* (Taylor & Francis Ltd, http://www.tandfonline.com), Volume 6, issue no. 2: pp. 81–87.
Study granted by NIKE Inc. USA.

References

Council of Europe. (1988). EUROFIT: Handbook for the EUROFIT tests of physical fitness. Council of Europe: Committee for the development of Sport, Rome, Italy.
De Clercq, D., Debuyck, G., Gerlo, J., Rambour, S., Segers, V. & Van Caekenberghe, I. (2014). Cutting performance wearing different studded socces shoes on dry and wet artificial turf. *Footwear Science*, 6 (2), 81–87.
FIFA. (2009). *FIFA Quality Concept for Football Turf, Handbook of requirements*, Zurich: FIFA.
FIFA Quality Programme. (2016). FIFA Recommended Pitches. Retrieved from http://quality.fifa.com/en. Accessed 24/12/2016.

Heidt, R. S., Dormer, S. G., Cawley, P. W., Scranton, P. E., Losse, G., & Howard, M. (1996). Differences in friction and torsional resistance in athletic shoe-turf surface interfaces. *The American Journal of Sports Medicine*, 24(6), 834.

Hennig, E.M., & Sterzing, T. (2010). The influence of soccer shoe design on playing performance: A series of biomechanical studies. *Footwear Science*, 2(1), 3–11.

Lake, M. (2000). Determining the protective function of sports footwear. *Ergonomics*, 43(10), 1610–1621.

McGhie, D., & Ettema, G. 2013. Biomechanical analysis of traction at the shoe–surface interface on third-generation artificial turf. *Sports Engineering* 16(2), 71–80.

Müller, C., Sterzing, T., Lake, M., & Milani, T. (2010). Different stud configurations cause movement adaptations during a soccer turning movement. *Footwear Science*, 2(1), 21–28.

Müller, C., et al., (2010). Comprehensive evaluation of player surface interaction on artificial soccer turf. *Sports Biomechanics*, 9(3), 193–205.

Potthast, W., Verhelst, R., Hughes, M., Stone, K., & De Clercq, D. (2010). Football-specific evaluation of player–surface interaction on different football turf systems. *Sports Technology*, 3(1), 5–12.

Severn, K. A., Flemming, P. R., & Dixon, N. (2008). Science of synthetic turf surfaces: Player interactions. *Proceedings of 7th ISEA Conference, 3–6 June 2008 Biarritz, France*, 125.

Shorten, M. R., Hudson, B., & Himmelsbach, J. (2003). Shoe-surface traction of conventional and in-filled synthetic turf football surfaces. In: P. Milburn (Ed.) *Proceedings of XIX Congress of the International Society of Biomechanics* (pp. 6–11), Dunedin: University of Otago.

Sterzing, T., Müller, C., Schwanitz, S., Odenwald, S. & Milani, T. L. (2008). Discrepancies between mechanical and biomechanical measurements of soccer shoe traction on artificial turf. In Y-H. Kwon, J. Shim, J. K. Shim & I-S. Shin (Eds.), *Proceedings of the 26th Symposium of the International Society of Biomechanics in Sports* (pp. 339–342). Seoul: Korean Society of Biomechanics & Seoul National University.

Sterzing, T., Müller, C., Hennig, E. M., & Milani, T. L. (2009). Actual and perceived running performance in soccer shoes: A series of eight studies. *Footwear Science*, 1(1), 5–17.

Sterzing, T., Müller, C., & Milani, T.L. (2010). Traction on artificial turf: development of a soccer outsole. *Footwear Science*, 2(1), 37–49.

Wannop, J. W., & Stefanyshyn, D.J. (2012). The effect of normal load, speed and moisture on footwear traction. *Footwear Science*, 4(1), 37–43.

Part IV
Skill acquisition and training

14 Technique modifications to create more powerful kicking actions in experienced players

Neal Smith and Simon Augustus

1. Introduction

As we have noted in the previous chapters of this book, the maximal instep kick is an important variation of the kicking skill in soccer, as it is the most commonly used technique when attempting a direct shot at goal. The ability to generate a fast ball velocity represents a distinct advantage for a player when shooting, as this gives goalkeepers less time to react and increases the chances of scoring (Kellis & Katis, 2007, Inoue *et al.*, 2014; Lees *et al.*, 2010). A detailed understanding of the mechanisms that determine kicking performance is therefore important to inform coaching practices. Subsequently, the kinetic (Dorge *et al.*, 2002; Inoue *et al.*, 2014; Lees *et al.*, 2009; Nunome *et al.*, 2002; Nunome, Ikehami *et al.*, 2006) kinematic (Apriantono *et al.*, 2006; Andersen *et al.*, 1999; Levanon & Dapena, 1998; Nunome, Lake *et al.*, 2006) and electromyographic (Dorge *et al.*, 1999; Katis *et al.*, 2013) characteristics of mature maximal instep kick technique have been extensively documented, and most of these areas have been explained in Part I of this textbook. However, these investigations have been mostly descriptive in nature and the practical applications are limited. Only a few studies have attempted to improve maximal instep kicking performance through resistance training programs (Manolopoulos *et al.*, 2013; Manolopoulos *et al.*, 2006) and to our knowledge no empirical investigations have attempted to refine kicking technique to improve performance. The theoretical aspects that apply to how we may go about altering an individual's technique are covered in Chapter 15, yet this has not yet been successfully applied to technique change in football to gain performance improvement.

Co-ordinated instep soccer kicking involves the controlled recruitment of muscular and motion-dependent (from segment interactions) joint torques and the proximal-to-distal motion of the kicking leg is well established (Nunome, Ikegami *et al.*, 2006; Putnam, 1991; Putnam, 1993). That is, the kicking leg acts as an open kinetic chain that rotates around the pelvis to maximize shank and foot velocities at ball contact (Dorge *et al.*, 2002; Nunome, Ikegami, *et al.*, 2006). Less attention has been paid to the function of the support leg with regards to kicking performance, despite evidence to suggest the proximal-to-distal sequencing of the kick emanates from support leg action. For example, it has been shown that players who produce largest kicking hip vertical displacement generate the fastest shank angular velocities at ball contact (Inoue *et al.*, 2000). That is, extension of the support leg knee and hip during the kicking stride serves to lift the kicking leg hip; creating a motion-dependent moment which accelerates the kicking leg shank during its downswing (Nunome & Ikegami, 2005). More recently, it has been established that the support leg may contribute to performance by lifting the body and adding to the vertical velocity of the foot at impact (Lees *et al.* 2009) and an increasing joint reaction moment on the support leg side may decelerate the support leg hip and emphasize the forward rotation of the pelvis about the support leg hip and thigh towards the ball (Inoue *et al.*, 2014).

Clearly a kinetic link exists between the kicking and support legs during the maximal instep kick, but exactly how the support leg interacts to facilitate the coordinated downswing of the kicking leg during the kicking stride is still largely unknown. The question also remains whether pronounced vertical displacement of the hips (via support leg action) can be intentionally utilized to facilitate a faster kicking leg swing. However, it is logical to surmise that larger vertical displacement of the hips might be indicative of increased kicking performance since robust relationships exist between a) support knee and hip extension and shank angular velocity at ball contact (Inoue *et al.*, 2000; Nunome & Ikegami, 2005), and b) shank angular velocity at ball contact and peak ball velocity (De Witt & Hinrichs, 2012; Levanon & Dapena, 1998). The aims of the current study were therefore to: a) assess the effectiveness of a Technique Refinement Intervention designed to produce pronounced extension of the support leg knee and vertical displacement of the kicking hip joint during the kicking stride; and b) highlight the dynamic interaction between support and kicking legs during the maximal instep kick. We hypothesized that kicking performance would improve (i.e. increased ball velocity) following the Intervention program.

2. Method

2.1 Participants

Nine skilled club players (age 23.7 ± 3.8 years, height 1.82 ± 0.06 m, body mass 78.5 ± 6.1 kg; mean $\pm$ SD) volunteered for the investigation. All were regularly competing in senior amateur or semi-professional competitions, had a minimum of ten years' playing experience (14.7 ± 3.8 years) and were free from injury at the time of testing. All participants preferred to kick with the right foot. Informed consent was obtained prior to testing and ethical approval granted by the University's Local Ethics Committee.

2.2 Experimental design

The participants performed 10 maximal instep kicks both prior to and immediately following the Technique Refinement Intervention (there is a Technique Refinement Intervention sub-section below). The first 10 trials were performed with the participant's normal kicking technique to establish a representative baseline of technique and performance (NORM). The 10 trials following the intervention were performed with the refined technique (INT). Ten trials were chosen per condition as 10–15 trials is optimal for reducing typical error (within-subject variation) for variables commonly used to describe maximal instep kicking (Lees & Rahnama, 2013).

2.3 Technique Refinement Intervention

The intervention aimed to produce pronounced extension of the support leg knee and hip and vertical displacement of the pelvis and hips during the kicking stride. The intervention incorporated aspects of Carson and Collin's (2011) Five-A model for technical refinement in skilled performers (see Table 14.1). The intervention was split into two distinct phases; an Awareness Phase and an Adjustment Phase. During the initial Awareness Phase, the aim was for the participant to call into consciousness the differences between NORM and INT techniques. The Adjustment Phase then aimed to modify the technique and internalize the changes to the extent that it was no longer in conscious awareness. Care was taken not to make specific reference to individual body segments or positions during the intervention process, because implicit learning techniques have been reported to be more effective than explicit techniques when refining well developed movement patterns (Carson & Collins, 2011; MacPherson, Collins & Obhi, 2009).

Intervention sessions lasted 2–4 hours, were semi-structured and an iterative process whereby participants could revisit the material provided during the Awareness Phase if required. All intervention sessions were led by the same investigator to ensure consistency in delivery and implementation of the techniques used and feedback provided. Self-report was chosen to assess when each participant's technique had been successfully adjusted as kinematic measures may not be indicative of performance when refining movement patterns (Peh *et al.*, 2011). However, as outlined in Table 14.1, the lead investigator did qualitatively assess if the desired changes were apparent in each participant's INT technique.

2.4 Data collection and processing

All kicks were performed in a carpeted laboratory with the participants' preferred (right) foot using a FIFA-approved size five ball (inflated pressure 800 hPa). After warm up, participants were instructed to strike the ball as forcefully as possible into the centre of a catching net placed four metres away and approached the ball in the way most comfortable to them for the two specific kick conditions. The ball was placed so that the support (left) foot landed on a Kistler 9821B force platform (Kistler Instruments, Hook, UK) which collected ground reaction forces at 1000 Hz. The force platform was synchronized electronically with a 10-camera opto-electronic motion analysis system (250 Hz) (Vicon T40S, Vicon Motion Systems, Oxford, UK). A Casio Exilim EX-FH20 (Casio Ltd, Tokyo, Japan) digital camera (210 Hz) was used to provide qualitative feedback during the intervention process. The participant wore their usual Astroturf or indoor soccer shoes and a compressive shirt, shorts and socks for all trials.

Table 14.1 Detailed overview of procedures and techniques implemented during the Technique Refinement Intervention. Modified from Augustus *et al.* (2016) in *Journal of Sports Sciences.*

Awareness Phase	
Procedure	*Techniques used (from Carson and Collins, 2011)*
1. Provided a brief overview and participants informed study aimed to refine their kicking technique. 2. Showed video clips of elite performers using the desired technique. Emphasis placed on a long final kicking stride and low to high translation of centre of mass and momentum throughout the kicking stride and follow through, resulting in both feet leaving the ground. 3. Visual 3D animation from a previous performer using the desired technique (same level of experience as participant) used to further highlight these points and for slow motion example.	• Contrast/Awareness drills. • Mental and physical contrast of the current followed by new technique, aided by video.
'Approach the ball with increasing step length, displace your body weight from low to high during the kicking stride, strike the ball as forcefully as possible and follow through fully, leaving the ground and landing again on the kicking leg'	• Continuous discussion with investigators as to the solution for new technique.

Adjustment Phase	
Procedure	*Techniques used (from Carson and Collins, 2011)*

1. Participant continues to practice and discover the refined technique.
2. Verbal feedback provided ad hoc by researcher in relation to cues.
3. Qualitative feedback provided using Casio Exilim® Digital camera (210Hz) and Quintic Biomechanics (v21 Quintic Consultancy Ltd, Sutton Coldfield, UK) to allow participant to further refine technique.
4. Global kicking cue presented:
 	
5. Participant self-rates each practice kick (1 being poorest and 10 being perfect) on three questions:
  a. How well do you think you produced the best possible ball contact?
  b. How well do you think you performed co-ordinated kicking motion?
  c. How well do you think you performed the kick in relation to 'cues' given beforehand?
 	
6. When participant was consistently scoring >8 on all three questions for 5 consecutive practice kicks and the researcher was confident the desired changes had been made successfully, the participant proceeded to perform the 10 intervention trials.
	• Contrast/awareness drills (NORM vs. INT).
	
	• Investigator and video feedback.
	
	• Introduction of a holistic rhythm-based cue.
	
	• Confirmatory video analysis.
	
	• Self-rating scale for performance of new technique.

Prior to data collection, 24 passive reflective markers (12.6 mm diameter) were attached to selected lower limb landmarks as shown in Figure 14.1. To reduce error associated with soft tissue artefact, marker clusters (consisting of three markers fixed to semi-rigid plastic) were attached to the left and right thigh and shank to determine the orientation of these segments relative to the calculated anatomical joint centres obtained following static calibration (Cappozzo *et al.*, 1996). One additional marker was cut into hemispheres and placed over opposing poles of the ball so that ball velocity could be calculated. Raw marker displacements were smoothed within the Vicon Nexus software (Vicon Nexus v1.8.2, Vicon Motion Systems, Oxford, UK) using a generalized, cross-validated spline (GCVSPL) (Woltring, 1986) (30 MSE; chosen as per residual analysis (Winter, 2009)). Due to distortions of position and velocity data associated with marker trajectories through impacts (Knudson & Bahamonde, 2001; Nunome, *et al.*, 2006), trajectories during the ball impact phase (one frame before and five after ball contact) were extrapolated using the same GCVSPL function.

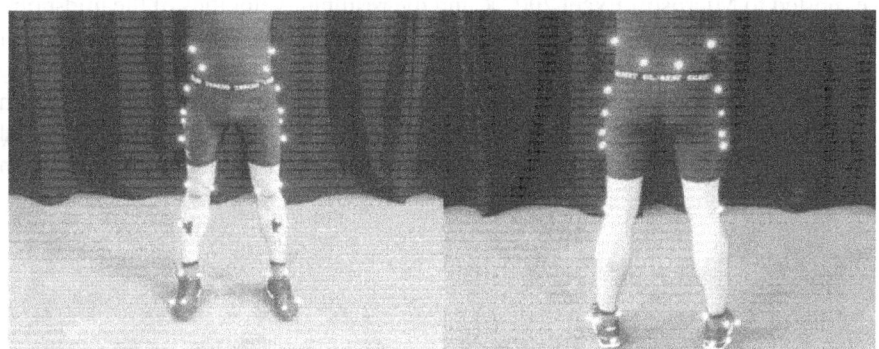

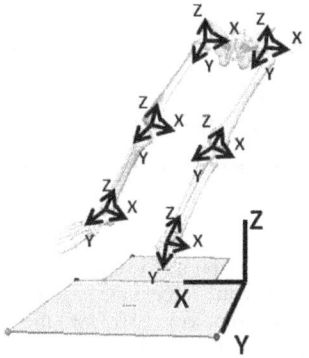

Figure 14.1 Reflective marker placements. The trochanter, femoral epicondyle, malleoli and kicking foot 2nd metatarsal markers were removed following static calibration (top panel). Definition of lab and joint co-ordinate systems. At each joint Z = interval/ external rotation, Y= abduction/ adduction and X = flexion/ extension (bottom panel). Modified from Augustus *et al.* (2016) in *Journal of Sports Sciences*

Synchronized force and 3D motion data were exported to Visual 3D (v5.00.31, C-Motion, Rockville, USA) where support and kicking leg knee and hip joint powers (generation/absorption), moments (flexion/extension), reaction forces (compressive/tensile) and angular velocities (flexion/extension) were calculated. Lower limb motion was defined using a seven segment, six degrees of freedom model including the pelvis, thighs, shanks and feet. Geometrical volumes were used to represent individual segments and inertial parameters were derived from young male Caucasians (De Leva, 1996). For all segments joint co-ordinate systems were defined at the proximal joint (see Figure 14.2), whereby hip joint centres were estimated from the positions of the pelvic markers (Bell *et al.*, 1989) and knee and ankle joint centres were defined as the mid-point between femoral epicondyle and malleoli marker, respectively. Joint angle orientations were defined by the distal joint segment relative to the proximal, using an X-Y-Z Cardan rotation sequence (Lees, Barton *et al.*, 2010). All kinetic data were resolved to the proximal co-ordinate system and were normalized to body mass. The smoothed co-ordinates of the ball markers were exported to Microsoft Excel 2007® and the resultant velocities of the mid-point between the two markers were computed at each frame following ball contact to ascertain the peak resultant ball velocity of each kicking trial. Kicking motions were time-normalized between the instances of support foot touchdown (SFTD) (0%) and ball contact (BC) (100%) and key events and phases defined as shown in Figure 14.2. For discrete measures, the average value from each

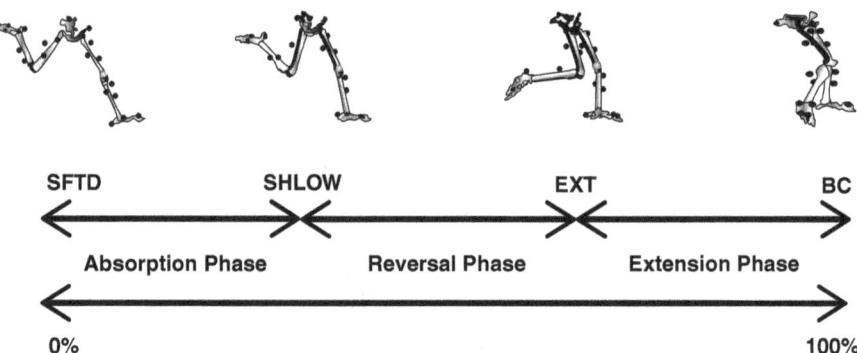

Figure 14.2 Definition of kicking motion key events and phases. Support foot touchdown (SFTD) was the instance the force plate began to measure a vertical force (25 N threshold), support hip joint low (SHLOW) the instance the calculated support hip joint centre was at its lowest displacement in the global Z (vertical) plane, support knee extension (EXT) the instance the support leg knee began to exhibit an extension angular velocity and ball contact (BC) was one frame before the ball markers showed a clear onset of forward movement. Subsequently, Absorption Phase occurred between SFTD and SHLOW, Reversal Phase between SHLOW and EXT and Extension Phase between EXT and BC. Modified from Augustus *et al.* (2016) in *Journal of Sports Sciences*

participant's 10 trials were used to calculate a group mean per condition. Whereas time-series data from all trials per participant were included to calculate a mean curve per condition. Thus, data are expressed as mean ± SD per condition.

2.5 Statistical analyses

To assess if the intervention process had successfully refined kicking technique two-tailed paired *t*-tests were conducted using SPSS (v20; SPSS Inc., Chicago, IL). These compared the peak ball velocities and vertical (Z axis) displacements of the kicking hip joint centre from support hip low (SHLOW) to ball contact (BC) between the two kicking conditions. Overall alpha was Bonferroni adjusted to α = 0.025 and effect sizes were calculated using Cohen's *d* (Cohen, 1988). To compare the time-normalized kinematic and kinetic waveforms, Statistical Parametric Mapping (SPM) was conducted using freely available source code (SPM1D v0.1, (Pataky, 2012)) in Python (Python v2.7.2; Enthought Python Distribution, Austin, USA). SPM allows for quantitative evaluation of differences across the entire kicking motion rather than at pre-selected discrete instances and removes the bias of analysing one-dimensional data using zero-dimensional (discrete) techniques (Pataky *et al.*, 2015). First, a paired *t*-test statistical curve (SPM{*t*}) was calculated for each dependent variable (Robinson *et al.*, 2014) across the entire kicking motion. Next, the significance of the SPM{*t*} supra-threshold clusters were determined topologically using random field theory (Adler & Taylor, 2009). Alpha was Bonferroni adjusted to α = 0.003 to account for multiple comparisons (N = 16). That is, where the SPM{*t*} curve exceeded the critical *t*-threshold at which only α% of smooth random curves would be expected to traverse, there was deemed to be a significant difference between conditions. Conceptually, a SPM paired *t*-test is therefore calculated and interpreted similarly to a scalar (discrete) paired *t*-test (Pataky, 2015).

3. Results

The peak ball velocities following INT (26.3 ± 2.1 m·s^{-1}) were significantly faster ($P < 0.025$) than those observed during the NORM trials (25.1 ± 1.5 m·s^{-1}). Vertical displacements of the calculated kicking leg hip joint centers from SHLOW to BC were significantly larger ($P < 0.025$) in the INT trials (0.041 ± 0.012 m) than in the NORM trials (0.028 ± 0.011 m). Table 14.2 shows detailed results of the paired *t*-tests. During the NORM condition the Absorption and Reversal Phases constituted $46 \pm 7\%$ and $34 \pm 7\%$ of total kicking motion, respectively; whereas these same phases lasted $41 \pm 7\%$ and $34 \pm 12\%$ when kicks were performed with the INT technique. The Extension Phase lasted $20 \pm 10\%$ during NORM compared to $25 \pm 7\%$ in the INT condition. SPM graphics provide a shaded darker portion (grey) where significant differences between the two mean curves occur.

Table 14.2 Paired *t*-test results comparing discrete measures of performance between the NORM and INT conditions. Modified from Augustus *et al*. (2016) in *Journal of Sports Sciences*.

	p-Value	*Mean Difference*	*Effect Size (Cohen's d**)*	*95% Confidence Interval*	
				Lower	Upper
Peak Ball Velocity (m·s⁻¹)	p < 0.001*	1.2 m·s⁻¹	0.58	0.7 m·s⁻¹	1.7 m·s⁻¹
Vertical Displacement of Kicking Hip Joint Centre (m)	p < 0.001*	0.012 m	0.89	0.009 m	0.015 m

* Denotes significant difference between INT and NORM conditions, P < 0.025.
** d = 0.2 – 0.5, small effect. d = 0.5 – 0.8, medium effect. d > 0.8, large effect.

3.1 Support leg

Figures 14.3 and 14.4 illustrate support leg joint profiles from the two conditions and subsequent statistical results. In the period immediately preceding ball contact (99%–100% of kicking motion) the support knee was extending significantly faster ($P < 0.003$) during the INT trials. The support knee moment observed during the period that corresponded with peak extension (12–17%) was significantly larger during the INT condition ($P < 0.003$). Similarly, compressive reaction forces at the support knee were significantly larger in the INT condition at 12–17%, 25–29% and from 49–100% of total kicking motion ($P < 0.003$). No significant differences in support knee power, or support hip extension angular velocity were observed ($P > 0.003$). However, support hip extension moment and compressive reaction forces were significantly larger between 12–17% and 10–16% of kicking motion during the INT trials, respectively ($P < 0.003$). Finally, support hip compressive reaction force was also significantly larger (43–100%, $P < 0.003$) and significantly more power was generated throughout the Reversal and Extension during the INT condition (52–100%, $P < 0.003$).

Kicking leg

Figures 14.5 and 14.6 illustrate kicking leg joint profiles from the two conditions and subsequent statistical results. Kicking hip flexion moment during the initial period of the Reversal Phase (45–60%) was significantly greater in the NORM condition ($P < 0.003$). Kicking hip tensile reaction force was significantly larger between 10–96% of total kicking motion when performed with the INT technique ($P < 0.003$). As the kicking motion progressed, the kicking hip generated less power and power absorption was noted in both conditions in the period immediately preceding BC (90–100% of kicking motion). During the latter part of the Reversal Phase and entire Extension Phase until BC (70–100%), the knee was extending at a significantly faster rate when kicks were performed with the INT technique ($P < 0.003$). After the kicking knee moment reversed at around

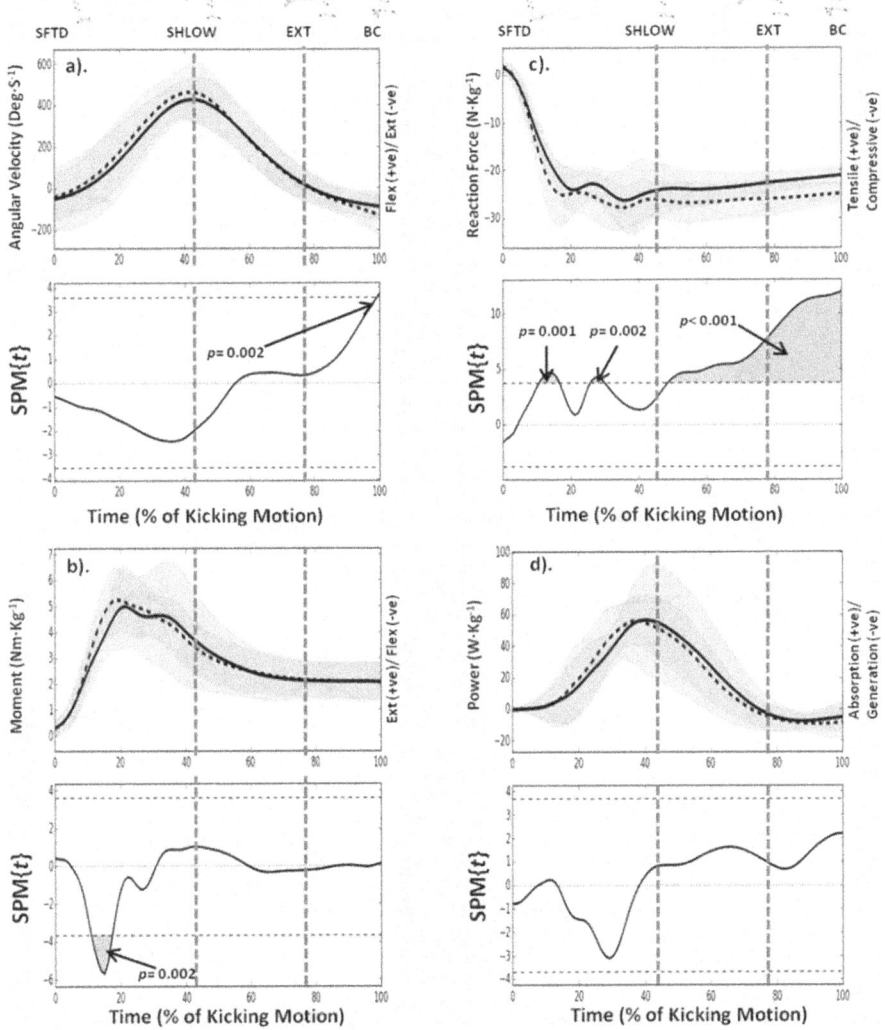

Figure 14.3 Mean ± SD support knee joint angular velocities (a), moments (b), reaction forces (c) and powers (d) observed during the NORM (bold) and INT (dashed) conditions between SFTD (0%) and BC (100%). Below each joint parameter curve is the corresponding SPM{*t*} output. Shaded areas and *p*-value labels indicate SPM{*t*} threshold (dotted horizontal line) has been exceeded and there is a significant difference between conditions (α = 0.003). Vertical dashed lines indicate average SHLOW and EXT events across all trials. Ext = Extension, Flex = Flexion. Modified from Augustus *et al.* (2016) in *Journal of Sports Sciences*

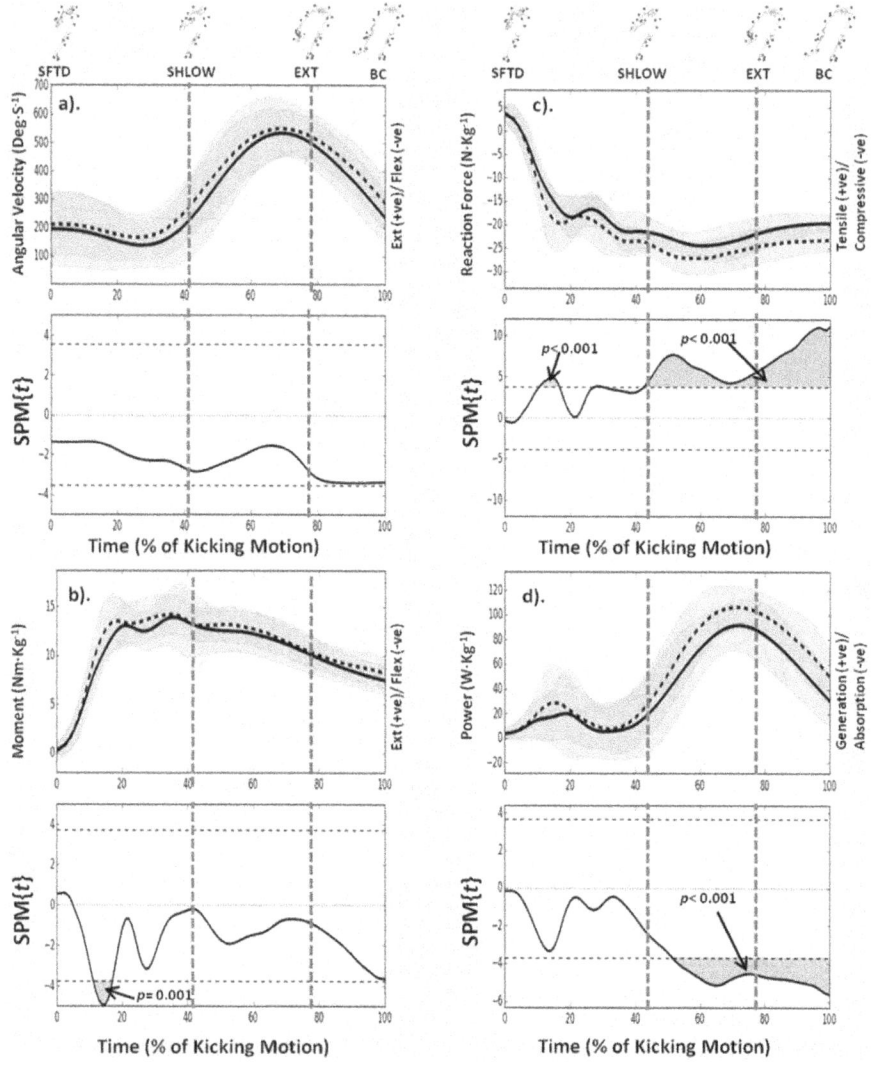

Figure 14.4 Mean ± SD support hip joint angular velocities (a), moments (b), reaction forces (c) and powers (d) observed during the NORM (bold) and INT (dashed) conditions between SFTD (0%) and BC (100%). Below each joint parameter curve is the corresponding SPM{t} output. Shaded areas and *p*-value labels indicate SPM{t} threshold (dotted horizontal line) has been exceeded and there is a significant difference between conditions ($\alpha = 0.003$). Vertical dashed lines indicate average SHLOW and EXT events across all trials. Ext = Extension, Flex = Flexion. Modified from Augustus *et al.* (2016) in *Journal of Sports Sciences*

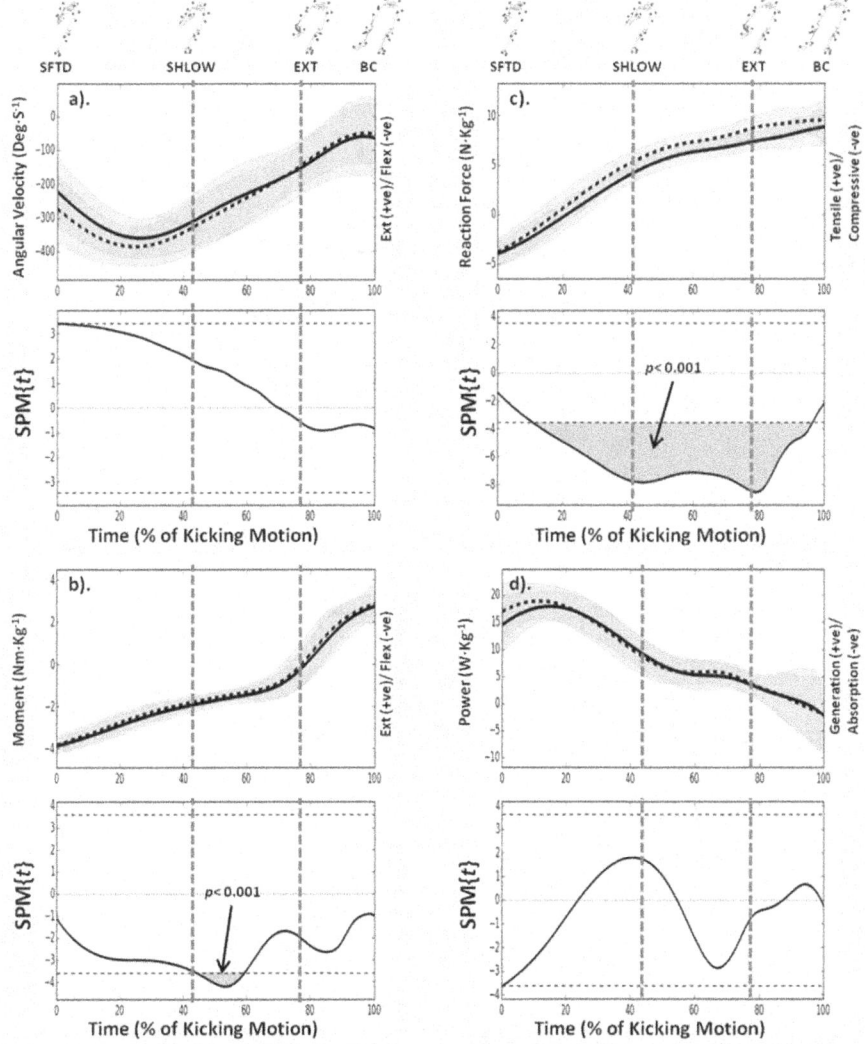

Figure 14.5 Mean ± SD kicking hip joint angular velocities (a), moments (b), reaction forces (c) and powers (d) observed during the NORM (bold) and INT (dashed) conditions between SFTD (0%) and BC (100%). Below each joint parameter curve is the corresponding SPM{*t*} output. Shaded areas and *p*-value labels indicate SPM{*t*} threshold (dotted horizontal line) has been exceeded and there is a significant difference between conditions ($\alpha = 0.003$). Vertical dashed lines indicate average SHLOW and EXT events across all trials. Ext = Extension, Flex = Flexion. Modified from Augustus *et al.* (2016) in *Journal of Sports Sciences*

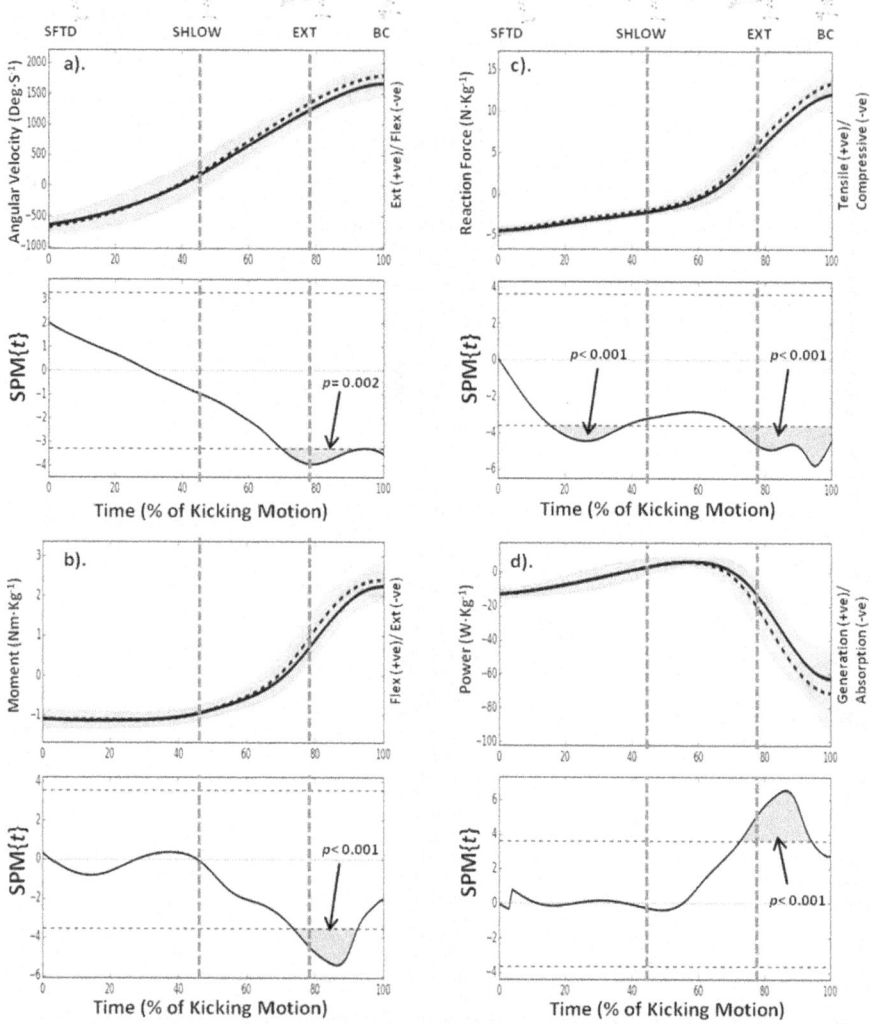

Figure 14.6 Mean ± SD kicking knee joint angular velocities (a), moments (b), reaction forces (c) and powers (d) observed during the NORM (bold) and INT (dashed) conditions between SFTD (0%) and BC (100%). Below each joint parameter curve is the corresponding SPM{*t*} output. Shaded areas and *p*-value labels indicate SPM{*t*} threshold (dotted horizontal line) has been exceeded and there is a significant difference between conditions (α = 0.003). Vertical dashed lines indicate average SHLOW and EXT events across all trials. Ext = Extension, Flex = Flexion. Modified from Augustus *et al.* (2016) in *Journal of Sports Sciences*

70% of total kicking motion the INT technique elicited a significantly larger flexion moment between 74–92% of the movement ($P < 0.003$). Similarly, a significantly larger tensile reaction force was seen when the kicks were performed with the INT technique from 70% of motion to BC ($P < 0.003$). Further, an expeditious increase in power absorption at the kicking knee is seen during the Extension Phase and power absorption is significantly larger when kicks are performed with the INT (72–93%) ($P < 0.003$).

4. Discussion

4.1 Effectiveness of Technique Refinement Intervention

Kicking performance was enhanced following the Technique Refinement Intervention since peak ball velocities and kicking knee angular extension velocities at BC were significantly faster during the INT condition. Furthermore, the Intervention successfully elicited significantly greater extension of the support leg knee and vertical displacement of the kicking hip joint during the kicking stride. As a robust relationship exists between ball velocity and the linear and angular velocities of the kicking foot at BC, it is widely considered that maximizing these two variables is integral to performance of the maximal instep kick (DeWitt & Hinrichs, 2012; Kellis & Katis, 2007; Nunome, Ikegami *et al.* 2006; Levanon & Dapena, 1998). Further, since the kicking ankle is forced into plantar-flexion during foot–ball impact (Nunome, Lake *et al.,* 2006) the knee is considered the most distal joint which can facilitate faster foot velocities at BC. However, a key caveat of this relationship is that ball velocity is also dependent on the quality of foot–ball impact (Andersen *et al.,* 1999; Nunome, Lake *et al.* 2006); thus increasing foot velocity at BC is not wholly indicative of performance. Indeed, re-organization of movement patterns can often lead to performance decrements due to 'collapse' of technique (Carson & Collins, 2011; MacPherson *et al.*, 2009). Had this been the case within the relatively short intervention period we speculate that it is likely foot–ball impact quality may have reduced, leading to a decrement in peak ball velocity. Conversely, we argue that because the alterations made to support leg action during the intervention process were subtle, the participants were able to produce significantly faster kicking knee extension velocities during the INT condition without compromising the dynamic stability and precise foot–ball impact mechanics needed for a successful kick (Lees *et al.*, 2009). Ultimately, the increase in kicking knee velocity observed at BC following intervention accounted for the concurrent increase in ball velocity; and as such our hypothesis that kicking performance would be improved was confirmed.

4.2 Contribution of support leg to performance

The greater support leg hip and knee extension in the final Extension Phase of the kicking stride during the INT condition served to lift the support leg hip vertically and promote the downward (extension) velocity of the knee towards the ball.

Previous studies have highlighted that the motion-dependent extension moment at the kicking knee due to vertical hip displacement as described by Putnam (1991) is greater when support leg hip vertical acceleration is larger (Inoue *et al.*, 2014; Nunome & Ikegami, 2005). However, neither study reported kicking leg kinematic data to support the conclusion that this mechanism directly influences leg swing speed. In the present study the instance of support knee extension (EXT) and power generation was coupled with the kicking knee's increase in power absorption, tensile reaction force and extension angular velocity through to BC; indicating that the kicking shank and foot was being accelerated passively about the knee towards the ball. Further, the kicking knee was showing a larger flexion moment during the Extension Phase of the INT trials which commonly occurs to protect the kicking knee joint as it is prepared for contact (Kellis & Katis, 2007; Lees *et al.*, 2009). A backwards (flexion) moment also supports the notion that the shank cannot be accelerated via muscular forces during the Extension Phase and the speeds of the kicking knee at BC certainly exceed the inherent force-velocity capabilities of the musculature (Nunome, Ikegami *et al.*, 2006). As such, the motion-dependent interaction between the kicking thigh and shank has been identified as the dominant action by which the shank is passively accelerated during the downswing (Dörge *et al.*, 2002; Nunome, Ikegami *et al.*, 2006). We argue however that it is not sufficient to illustrate the dynamics of maximal kicking performance using data from the kicking leg only, since kinetic sources originating from support leg action directly contribute to shank angular velocity during the Extension Phase. That is, when kicks were performed following the INT passive contribution to shank acceleration was exacerbated since kicking knee power absorption, tensile reaction forces and extension angular velocities were significantly larger throughout most of the Extension Phase. Thus the assumptions made previously regarding the relationship between vertical hip acceleration and passive acceleration of the shank before BC (Inoue *et al.*, 2014; Nunome & Ikegami, 2005) are confirmed. However, because the pronounced passive contribution to kicking shank extension during the INT condition begins before EXT and support knee extension, velocity is only faster for a brief period before BC other kinetic sources originating from the support leg may also influence kicking leg velocity during the downswing.

The ability of the support leg knee and hip to contribute to performance during the final Extension Phase might originate from the dynamics that occurred during the preceding phases. It is well established that the support knee joint is forced into flexion following SFTD to dissipate ground reaction forces (GRFs) and a large counteracting (extension) knee moment resists this flexion to ensure the body is kept stable through the movement (Inoue *et al.*, 2014; Lees *et al.*, 2009). This large extension moment is replicated in the current study irrespective of condition; but following INT, participants exhibited significantly larger peak moments and compressive reaction forces at the support knee and hip during the Reversal Phase. This suggests that participants were actively contracting the support knee extensor musculature to resist knee flexion following SFTD and thus performed the movement with a more rigid support leg (Inoue *et al.*, 2014).

One benefit of actively resisting flexion may be that the support leg is able to reverse from power absorption to generation sooner in the kicking motion, maximizing its potential to extend and contribute to performance in the latter phases of the kick. Indeed, the Absorption Phase duration was shorter during the INT compared to the NORM condition and the final Extension Phase was longer when performed with the INT condition.

Another benefit of actively resisting flexion following SFTD may be to minimize negative work and power absorption at the support knee to promote transfer of power through the support leg in a distal to proximal direction (i.e. from the ground to the support hip and pelvis). Indeed, compressive reaction forces at the support hip and knee were significantly larger during the INT trials for the duration of the Reversal and Extension Phases and the support hip was able to generate significantly more power during these phases following the INT. Inoue *et al.* (2014) previously noted that horizontal deceleration of the support leg hip and a large joint reaction force at the support hip following SFTD prompted the counter clockwise rotation of the pelvis about the support leg that precedes the proximal-to-distal sequencing of the kicking leg (Dörge *et al.*, 2002; Nunome, Ikegami *et al.*, 2006). However, despite presenting a more precise illustration of the dynamics interaction between the support leg and pelvis than shown here, the previous authors did not attempt to highlight how this interaction influenced kicking leg dynamics. In the present study kicking hip tensile reaction forces were significantly larger for the majority of the kicking motion (11–97%) when performed with the INT technique suggesting that the enhanced propagation of power through the closed kinetic chain of the support leg is translated across the pelvis into the open kinetic chain of the kicking leg. Further, because the greater passive power flow and extension velocity of the kicking knee observed during the INT condition occurs before the EXT event and support knee extension is only larger during the final 2% of kicking motion, kinetic sources other than the motion-dependent moment due to hip vertical acceleration (Inoue *et al.*, 2014; Nunome & Ikegami, 2005) must have been contributing to the acceleration of kicking knee, shank and foot towards BC.

The current study provides preliminary evidence for the application of Technique Refinement in skilled soccer players to enhance kicking performance, however its limitations must also be considered. First, the absence of a control (sham) training group should be noted. Had a paired group been included which received non-specific instruction during the Intervention process (i.e. not focused on increasing vertical hip displacement), we could be more confident that performance improvements were a result of the intervention process and the mechanisms presented rather than learning effects. Second, only the immediate effect of the Intervention was measured, thus further study is needed to examine its longitudinal applications. Specifically, the present data provides preliminary support for use of the 'Awareness' and 'Adjustment' aspects of the Five-A Model (Carson & Collins, 2011) for technical refinement of kicking but it is not known whether subsequent 'Automation' and 'Assurance' aspects can be incorporated as part of a more extensive intervention process. This would form a key inclusion for any further interventions on technique change, as currently we are confident our externally focused instruction programme provided performance improvement, we cannot be certain that the players' technique was fundamentally 'changed' unless a follow-up session was conducted to verify this. Finally, due to the experimental nature of the study, no accuracy or situational constraints (e.g. moving ball, opposing players) were introduced to the kicking task. Thus the findings are currently limited to 'set-piece' situations where production of a fast ball velocity is the main goal of the kick.

5. Conclusion

Preliminary evidence is presented to suggest that maximal instep kick technique can be refined through coaching interventions to elicit enhanced performance (i.e. faster ball velocity). Greater active contraction and extension of the support leg musculature during the kicking stride may facilitate power flow across the pelvis and passive acceleration of the lower leg to maximize foot linear and angular velocities at ball impact. This knowledge might influence coaching practices by: a) providing a basis from which to generate effective kicking interventions and b) highlighting the benefits of strengthening the support leg when training to improve kicking performance. Further, since support leg action can alter lower limb dynamics during kicking and contribute significantly to performance, it is not sufficient to illustrate the dynamics of maximal kicking using data obtained exclusively from the kicking leg.

References

Adler, R. J. & Taylor, J. E. (2009). *Random Fields and Geomtery*. New York, NY: Springer.

Andersen, T., Dorge, H. C., & Thomsen, F. (1999). Collisions in soccer kicking. *Sports Engineering*, 2, 121–125.

Apriantono, T., Nunome, H., Ikegami, Y., & Sano, S. (2006). The effect of muscle fatigue on instep kicking kinetics and kinematics in association football. *Journal of Sports Sciences*, 24, 951–960.

Augustus, S., Mundy, P. & Smith, N. (2016): Support leg action can contribute to maximal instep soccer kick performance: An intervention study. *Journal of Sports Sciences*, doi: 10.1080/02640414.2016.1156728

Bell, A. L., Brand, R. A., & Pedersen, D. R. (1989). Prediction of hip joint centre location from external landmarks. *Human Movement Science*, 8, 3–16.

Cappozzo, A., Catani, F., Leardini, A., Benedetti, M. G., & Della Croce, U. (1996). Position and orientation in space of bones during movement: Experimental artefacts. *Clinical Biomechanics*, 11, 90–100.

Carson, H. J., & Collins, D. (2011). Refining and regaining skills in fixation/diversification stage performers: The Five-A Model. *International Review of Sport and Exercise Psychology*, 4, 146–167.

Cohen, J. (1988). *Statistical Power Analysis for the Behavioral Sciences* (2nd ed.). New Jersey: Lawrence Erlbaum.

De Leva, P. (1996). Adjustments to Zatsiorsky-Seluyanov's segment inertia parameters. *Journal of Biomechanics*, 29, 1223–1230.

De Witt, J. K., & Hinrichs, R. N. (2012). Mechanical factors associated with the development of high ball velocity during an instep soccer kick. *Sports Biomechanics*, 11, 382–390.

Dorge, H. C., Andersen, T. B., Sorensen, H., & Simonsen, E. B. (2002). Biomechanical differences in soccer kicking with the preferred and the non-preferred leg. *Journal of Sports Sciences*, 20, 293–299.

Dörge, H. C., Andersen, T. B., Sorensen, H., Simonsen, E. B., Aagaard, H., Dyhre-Poulsen, P., & Klausen, K. (1999). EMG activity of the iliopsoas muscle and leg kinetics during the soccer place kick. *Scandinavian Journal of Medicine & Science in Sports*, 9, 195–200.

Inoue, S. I., Ito, T., Sueyoshi, Y., O'Donoghue, R. K., & Mochinaga, M. (2000). The effect of lifting the rotational axis on swing speed of the instep kick in soccer. In Y. Hong & D. P. Johns (Eds.), *Proceedings of 18th International Symposium on Biomechanics in Sports* (pp. 39–42). Hong Kong: The Chinese University of Hong Kong.

Inoue, K., Nunome, H., Sterzing, T., Shinkai, H., & Ikegami, Y. (2014). Dynamics of the support leg in soccer instep kicking. *Journal of Sports Sciences*, 32, 1023–1032.

Katis, A., Giannadakis, E., Kannas, T., Amiridis, I., Kellis, E., & Lees, A. (2013). Mechanisms that influence accuracy of the soccer kick. *Journal of Electromyography and Kinesiology*, 23, 125–131.

Kellis, E. & Katis, A. (2007). Biomechanical characteristics and determinants of instep soccer kick. *Journal of Sports Science and Medicine*, 6, 154–165.

Knudson, D., & Bahamonde, R. (2001). Effect of endpoint conditions on position and velocity near impact in tennis. *Journal of Sports Sciences*, 19(11), 839–844.

Lees, A., Asai, T., Andersen, T. B., Nunome, H., & Sterzing T. (2010). The biomechanics of kicking in soccer: A review. *Journal of Sports Sciences*, 28(8), 85–817.

Lees, A., Barton, G., & Robinson, M. (2010). The influence of Cardan rotation sequence on angular orientation data for the lower limb in the soccer kick. *Journal of Sports Sciences*, 28, 445–450.

Lees, A., & Rahnama, N. (2013). Variability and typical error in the kinematics and kinetics of the maximal instep kick in soccer. *Sports Biomechanics*, 12, 283–292.

Lees, A., Steward, I., Rahnama, N., & Barton, G. (2009). Lower limb function in the maximal instep kick in soccer. In T. Reilly & G. Atkinson (Eds.), *Contemporary Sport, Leisure and Ergonomics* (pp. 149–160). London: Routledge.

Levanon, J., & Dapena, J. (1998). Comparison of the kinematics of the full-instep and pass kicks in soccer. *Medicine and Science in Sports and Exercise*, 30, 917–927.

MacPherson, A. C., Collins, D., & Obhi, S. S. (2009). The importance of temporal structure and rhythm for the optimum performance of motor skills: A new focus for practitioners of sport psychology. *Journal of Applied Sport Psychology*, 21, S48–S61.

Manolopoulos, E., Papadopoulos, C., & Kellis, E. (2006). Effects of combined strength and kick coordination training on soccer kick biomechanics in amateur players. *Scandinavian Journal of Medicine & Science in Sports*, 16, 102–110.

Manolopoulos, E., Katis, A., Manolopoulos, K., Kalapotharakos, V., & Kellis, E. (2013). Effects of a 10-week resistance exercise program on soccer kick biomechanics and muscle strength. *The Journal of Strength & Conditioning Research*, 27, 3391–3401.

Nunome, H., Asai, T., Ikegami, Y., & Sakurai, S. (2002). Three-dimensional kinetic analysis of side-foot and instep soccer kicks. *Medicine and Science in Sports and Exercise*, 34, 2028–2036.

Nunome, H. & Ikegami, Y. (2005). The effect of hip linear motion on lower leg angular velocity during soccer instep kicking. In Q. Wang (Ed.) *Proceedings of the 23rd Symposium on Biomechanics in Sports* (pp. 770–772), Beijing: The China Institute of Sports Science.

Nunome, H., Ikegami, Y., Kozakia, R., Apriantono, T., & Sano, S. (2006). Segmental dynamics of soccer instep kick with the preferred and non-preferred leg. *Journal of Sports Sciences*, 24, 529–541.

Nunome, H., Lake, M., Georgakis, A., & Stergioulas, L. K. (2006). Impact phase kinematics of instep kicking in soccer. *Journal of Sports Sciences*, 24, 11–22.

Pataky, T. C. (2012). One-dimensional statistical parametric mapping in Python. *Computer Methods in Biomechanics and Biomedical Engineering*. 15, 295–301.

Pataky, T. C., Vanrenterghem, J., & Robinson, M. A. (2015). Zero- vs. one-dimensional, parametric vs. non-parametric, and confidence interval vs. hypothesis testing procedures in one-dimensional biomechanical trajectory analysis. *Journal of Biomechanics*, 48, 1277–1285.

Peh, S. Y. C., Chow, J. Y., & Davids, K. (2011). Focus of attention and its impact on movement behaviour. *Journal of Science and Medicine in Sport*, 14, 70–78.

Putnam, C. A. (1991). A segment interaction analysis of proximal-to-distal sequential segment motion patterns, *Medicine and Science in Sports and Exercise*, 23, 130–144.

Putnam, C. A. (1993). Sequential motions of body segments in striking and throwing skills: descriptions and explanations. *Journal of Biomechanics*, 26, 125–135.

Robinson, M. A., Donnelly, C. J., Tsao, J., & Vanrenterghem, J. (2013). Impact of knee modeling approach on indicators and classification of ACL injury risk. *Medicine and Science in Sports and Exercise*, 46, 1269–1276.

Selbie, W. S., Hamill, J., & Kepple, T. M. (2014) Three-dimensional kinetics. In *Research Methods in Biomechanics* (pp. 151–176). Champaign, IL: Human Kinetics.

Winter, D. A. (2009). *Biomechanics and Motor Control of Human Movement* (4th edn). Chichester: Wiley & Sons.

Woltring, H. J. (1986). A FORTRAN package for generalized, cross-validatory spline smoothing and differentiation. *Advances in Engineering Software*, 8, 104–113.

15 Skill change in elite-level kickers

Interdisciplinary considerations of an applied framework

Howie J. Carson, Dave Collins and Phil Kearney

1. Introduction

In coaching practice, technical preparation plays an important role. Therefore, interdisciplinary models which provide concrete starting-points for the improvement of technique are substantial for practical work. Coaches ... would like to know how to stimulate stable modes of coordination in the athlete, how to stabilize proper techniques, and how to change previously acquired, inefficient movement patterns during training. All these questions cannot be answered merely through biomechanical analyses or through detailed movement observations. In this context, relevant methods are rather those which comprehend and illuminate the cognitive–coordinative background of technique execution. (Schack & Bar-Eli, 2007, p. 63)

This quotation which spans a number of characteristics and elements of optimum high-performance environments, stresses the need for a cross- and interdisciplinary approach to support practice. Two important considerations are central to the pursuit of this ideal. Firstly, and in contrast to evidence-based guidance on stimulating (i.e., acquiring) and stabilising (i.e., performing) skills, there is a relative dearth when addressing how to *change* an athlete's already acquired and well-established movement (Carson & Collins, 2016; Fitts & Posner, 1967). Specifically, we refer here to such change as making a small tweak, or refinement, to technique in a way that is new to the athlete, although not of sufficient scope so as to constitute the complete acquisition of an entire new skill (cf. Carson & Collins, 2011). This scarcity of advice is unfortunate, since sporting domains present many situations when a technical refinement can generate significant performance improvements (e.g., Carson *et al.*, 2014; Hanin *et al.*, 2002). Examples of such situations include, when executing on new playing surfaces or with different equipment, responding to new playing styles of competitors, following the different challenges and styles posed by a new manager, or returning from injury. Crucially, coaches need to know how to implement refinement in a way that (1) changes remain permanent in the long term, and (2) ensures that the new version is robust against negative anxiety effects. These outcomes, or a lack thereof, are most clearly evident during closed and self-paced skills, when immense pressure during execution is loaded onto a single individual. As such, this chapter will directly focus on refining skills of this nature (e.g., penalty kicking). Indeed, anecdotal evidence has shown that considerable difficulty is experienced when attempting to realise these outcomes within professional team sports, such as penalty taking in rugby, soccer or hockey; perhaps as a consequence of employing coaching knowledge and techniques intended for different outcomes (i.e., acquisition versus present performance). Either way, there is a clear and current need within both academic and applied communities to understand why theory to explain skill acquisition and performance cannot be directly applied to athletes seeking long-term permanent and pressure resistant refinement.

Secondly, we welcome Schack and Bar-Eli's (2007) consideration of the oversight of not coaching both cognitive and co-ordinative aspects of skill execution. Indeed, while it should be obvious to readers that execution outcome is a direct result of kinematic and kinetic processes, a wealth of evidence has also demonstrated the perils of maladaptive conscious processing over these factors during highly stressful situations (e.g., Collins *et al.*, 2002; Hill & Shaw, 2013); a factor that must be proactively addressed if the skill is to be suitably 'pressure proof'. Equally important, however, is the athlete's attitude and intention to bring about change (Ajzen, 1991); therefore highlighting a breadth of cognitive factors that must also be catered for. Finally, and extending this interdisciplinary perspective, social factors can also be seen to significantly impact on a diverse range of outcomes during technical refinement, from programme adherence to the presentation/interpretation of feedback provided. Accordingly, it is insufficient and, in fact, misleading, to conceptualise optimum applied coaching solutions as being anything but biopsychosocial in approach (Bailey *et al.*, 2010; Collins *et al.*, 2012).

Taken together, this chapter presents a change in emphasis from solely addressing mechanical aspects of football movements (an important but inherently limited consideration), to understanding how the biomechanist can usefully support and act within an interdisciplinary coaching team to bring about effective refinement of closed and self-paced skills. Accordingly, we begin by addressing such team dynamics and the important underpinnings for successful co-operation across support practitioners. Following this, we outline and provide exemplification of a five-stage process, the Five-A Model (Carson & Collins, 2011), that is designed to facilitate the dual outcomes of long-term permanent and pressure proofed refinement. Finally, and highlighting the most significant contribution offered by the biomechanist, measures will be presented for both assessing and tracking athlete progress.

2. The importance of shared mental models

In supporting an athlete to undertake the difficult challenge of skill refinement, notable benefit may be found through various contributions from a range of support practitioners; namely, a coach, biomechanist, sport psychologist and skill development specialist, with these roles provided through either different specialists or multiple expertise individuals. Although one most typically considers the athlete–coach relationship as paramount (and there is no doubt that this is important), we propose that, in the present context, positive collaboration and input between practitioners is worth equal consideration (Burwitz *et al.*, 1994). Indeed, the impact of support team interaction has already been documented in other aspects of coaching practice; for example, in elite team culture change and talent selection panels. Underpinning such efficacy is the presence of a Shared Mental Model (SMM), or representation, of the task as a reference for coherent and reliable decision making (Collins & Hill, 2016). In this regard, the construction of a SMM should be derived from a Professional Judgement and Decision Making (PJDM) approach (Abraham & Collins, 2011; Collins *et al.*, 2015; Martindale & Collins, 2005), whereby multiple courses of action are generated and evaluated against planned outcomes and expected lines of consequence.

The application of a SMM has several important implications for the professional preparation, accreditation and practice of biomechanists. Indeed, many of these considerations are pan-domain, carrying similar consequences for other disciplines as for the constrained world of this (arguably) most objective of sport science disciplines. For the present discussion, however, we will focus on two: role clarity and the team dynamic.

Firstly, we look at the need for all concerned to understand and adhere to their role within the support process. With multidisciplinary support teams becoming the norm across high performance sport, and certainly a consistent feature of sports institutes worldwide, it is increasingly important to know exactly 'who does what, with whom and when'. This challenge was highlighted by Collins and Collins (2011) in considering the various roles, and potential for conflict, between the sports physician, physiotherapist, Strength and Conditioning (S and C) provider and coach. In simple terms, which specialist is responsible for communicating with the athlete? If they are all responsible, then the risk for mixed messages, confusion, angst and obfuscation is significant. Furthermore, how do the team communicate internally, to conduct the essential internal debate on how data are weighted and what actions are optimum? This situation is further complicated by the various phases through which support may be provided; for example, through the passage of an injury–diagnosis–rehabilitation–return to play cycle. The solution developed by the interdisciplinary team at UK Athletics (with full acknowledgement to Dr Bruce Hamilton and then Lead Physiotherapist Neil Black) was to phase support through these contexts, with each phase 'led' by the most appropriate disciplinary specialist. Thus, diagnosis was down to the doctor, early and mid-stage rehabilitation to the physiotherapist, later stage rehabilitation to the S and C provider, then return to play to the physiotherapist, S and C provider and coach. One of the most important principles here was that only the lead for that phase would communicate results and actions to the athlete–client. All others involved would feed data and actions through him/her, or at the very least ensure that everything was approved before they spoke one-on-one with the client! As a consequence, although debate behind the scenes/under the surface might be rigorous and vigorous, so far as the athlete–client was concerned, all was certainty, consistency and clarity (see Collins & Collins, 2011, for a more complete treatment of this approach).

Now consider the parallel situation for a biomechanist, hired to examine and evaluate the kicking performance of a senior player in a professional setup. Who is the client in this situation and, therefore, the target for feedback? The player or the coach? Then, in terms of action, who will now decide on, then direct, the actions taken as a result of the evaluation? Who will decide on the timing of this? Does the coach fully understand the time and resource implications of the refinement process which might be implicated (cf. the next section)? In short, who does the biomechanist tell, what, when and with what implications? Hopefully, this series of questions, both common and complex in our experience, offers a grounding in reality for the challenge of providing effective performance support.

The second, and related issue here is based around the team dynamic, and the 'rules' applied to their role execution by the different members of the support team (cf. Collins *et al.*, 2002). Returning to the rehabilitation example cited above, our original motivation for implementing the role clarity structure was because of obvious differences between specialists being aired to the athlete and coach, with predictably messy results! So, when physiotherapists and S and C providers would suggest diagnostic tests and checks, doctors suggest exercises, S and C providers suggest electro-treatment modalities and all offer differing views on the prognosis and pathway of return; the fans were clogged to say the least. In our skill refinement example, does the biomechanist 'merely' report the data and retire immediately? Does s/he offer implications of the data, together with various options? Indeed, is s/he trained to design interventions for such refinement? And even if they are, does this training extend to acknowledging and catering for the psycho-emotional implications and ensuring that the eventual execution is sufficiently pressure proof to withstand a clutch competition? In short, there is a lot to this.

Thus, the bottom line of all these concerns is that the team must 'enjoy' a dynamic through which differences can be aired and solved, with no negative implications for the client, be it athlete or coach. This is reflected in our statement made elsewhere, that high performing environments are characterised, or even classified, by the quality of disagreement. As Burke (2011) points out, high-performing teams must be comfortable with, even committed to, living life in the ZOUD...the Zone of Uncomfortable Debate.

In summary, the structure, role and dynamic between staff of every persuasion must be addressed and catered for, in an optimised, high-performing environment. That these considerations apply to even the most objective of disciplines represents an important realisation for the aspirant professional analyst. Consequently, the informed and aware biomechanist must avoid the simple trap of "I can measure it, so it's important and so is my advice" (cf. Collins *et al.*, 2015).

3. Implementing technical refinement

As established in the previous section, it is important that all members of the support team understand their role as part of a synergetic group. To assist in the development of an SMM, we now present the Five-A Model (Carson &

Collins, 2011). Specifically, this five-stage process is designed to deliver long-term permanent and pressure proof technical refinement through a variety of biomechanical, psychological and coaching 'tools'. In contrast to the mechanisms of acquiring (i.e., establishing technique and levels of automaticity) or prompting optimal performance of a skill (i.e., exploiting the already acquired technique and associated levels of automaticity), the Five-A Model explains refinement as requiring conscious deautomation of the aspect in need of modification (hereafter termed target variable), adjustment to a desired new version, reautomation of the new kinematics within the entire skill and finally, pressure proofing the skill to be 'competition ready'. Notably, due to our current emphasis on support practitioners operating in concert with others, the biomechanist (from this chapter's perspective) must appreciate and incorporate at least some elements of other disciplinary practice as a complimentary aspect; in short, knowing only about biomechanics is insufficient when seeking application within the coaching context.

3.1 Stage 1: analysis

Before deciding to refine technique, a detailed analysis must be undertaken by the interdisciplinary team. Indeed, for advanced athletes, technical refinement is an inevitably risky transition, unless it is known that technical refinement is necessary, what needs refining, how to refine and that a time has been identified for when refinement is appropriate. What the team must avoid is an athlete becoming trapped in a prolonged cyclical process of relapse and recycling through the refinement stages (typically observed during cessation of smoking habits; cf. Schachter, 1982). Accordingly, a PJDM approach to assessing the many factors required to answer these questions offers a sound starting point.

In determining the likelihood of successful refinement, considering the athlete's commitment and capability of achieving the course of action decided upon is paramount. To boost commitment levels, ideas and procedures may need to be 'sold' which, depending on the athlete's previous experience, intention (Ajzen, 1991) and reason for changing, could take up to several months, resulting in a decision to defer any potential refinement. Reflecting our earlier discussion of team dynamics, who does the selling must be carefully calculated based on the level of trust with the athlete. Of course, good coaching practice will have already equipped athletes with essential skills required to overcome such challenges/ transitions. As our previous work has identified, exposure or (better still) mastery of several psycho-behavioural characteristics during skill acquisition, can facilitate progress through what is an inevitably 'rocky road' (e.g., imagery, goal setting, motivation, vision of what it takes to succeed, social skills; MacNamara *et al.*, 2010; MacNamara *et al.*, 2008). For the biomechanist, similar 'pre-exposure' may require the athlete to feel confident in their ability to understand basic kinematic feedback (not so detailed as to cajole the athlete, or coach for that matter!) for use when evaluating training goals and a degree of familiarity with motion capture procedures and equipment; those too will have to be determined by the biomechanist as most appropriate. For instance, consider an athlete's willingness and ability to resist distraction from such equipment. Contrasting attitudes between an enthusiastic "I can really see that this is going to help me develop my skills" athlete and another "this is obtrusive and I'm not getting any better already" sceptical athlete, can seriously impact on the team's decision to even commence with the refinement intervention until this is rectified. That is not to say it is entirely the athlete's responsibility to adapt to the situation, however. Simplifying the process as much as it can be, may mean sacrificing some kinematic data and only recording the most essential elements of the skill. For example, our previous research (e.g., Carson & Collins, 2015) limited the analysis of golfers to upper body segments (pelvis upwards). From a pragmatic perspective, we employed mobile inertial measurement units over and above the 'reference standard' optoelectronic camera systems (e.g., Qualisys); our rationale being to ensure improved anatomical meaning (i.e., 3D data using local co-ordinate systems) versus the more usual video (i.e., 2D data from global co-ordinate systems) recording by a coach and, ecological validity when compared to indoor laboratory constraints. As such, it is important to recognise the various trade-off decisions that might need to be made when it comes to deviating from most typical anatomical modelling techniques and sampling rates, for example, to ensure all-important athlete 'buy-in'.

When assessing the case either for or against refinement, the necessity for, and technical aspect in need of, change must be on an individual basis; that is, avoiding the trap of Hume's Law (e.g., "Jonnie Wilkinson does that, so therefore Athlete 'X' ought to as well"). Once a skill has been learnt it is clear, simply from behavioural observation, that kickers demonstrate their own style of kicking (some technical aspects being more, others less, similar across individuals). What must be determined is whether these technical idiosyncrasies are 'errors' or in fact causative of successful executions? If the biomechanist is not well acquainted with the particular athlete's playing style, team role and technical capabilities, coach-guidance will be essential in translating what would ideally be a six-degrees-of-freedom analysis into technical principles that are widely used by athletes and coaches. Failure to establish even a general qualitative idea about potential target variables from those working closest to the athlete can, with tremendous frustration, lead to the situation of 'trying to find a needle in a haystack'. In summary, cross-team consensus, developed through triangulation across practitioners, is an important if sometimes illusive precursor for effective progress.

3.2 Stage 2: awareness

Having decided to implement technical refinement, the support team must now encourage increased conscious control over the flawed target variable. Indeed, it is widely argued that permanent change to an automated movement requires, at least temporarily, deautomation of the motor memory trace (Beilock *et al.*, 2002; Carson *et al.*, 2016; Oudejans *et al.*, 2007). In our previous research (Carson & Collins, 2015; Carson *et al.*, 2014; Collins *et al.*, 1999), and that of others (Hanin *et al.*, 2002; Hanin *et al.*, 2004), contrast drills have shown to be an effective coaching technique to help direct an athlete's attention narrowly inward (cf. Wulf, 2013). Specifically, these contrast drills require the athlete to perform alternate 'versions' of the skill, one that is 'correct/new' and one 'incorrect/old'. For example, Hanin *et al.* (2004) asked an Olympic swimmer to contrast the diving start position, height of jump and hand position involved during deep (old/incorrect technique) versus shallow (new/correct) water entries. In addition to this more conventional contrast, Collins *et al.* (1999) showed good effect when asking an Olympic javelin thrower to execute both left- and right-handed throws, simply to force greater concentration within the athlete. Undoubtedly, the ability of an athlete to forfeit subconscious control requires a good mental imagery ability, both in terms of visualisation and kinaesthetic acuity, to know how these two versions should be performed. Direct questioning (e.g., "tell me how it was different?") with the athlete to generate verbal 'cues' can help to clarify the motoric differences, since verbal and sensory memories are stored in parallel (i.e., one may activate the other; Paivio, 1986).

As expected, regaining conscious awareness can be very disorientating for the athlete and frustration can easily mount as performance drops in response to a regression in automaticity. Accordingly, manipulating the training environment can facilitate a most productive change in focus. Godbout and Boyd (2010) used a slower and more upright skating stride to allow an athlete to better sense the contrast in their ankle extension pattern during the cross-over skill. Collins *et al.* (1999) employed a shortened run-up during javelin throwing and Carson *et al.* (2016) a net for golfers in which to execute their shots. All of these less than 'representative' environments/tasks (cf. Pinder *et al.*, 2011) are intended to assist in initiating the awareness process by reducing influence from additional distractions. Therefore, optimum procedure includes highlighting skill refinement as a complex nonlinear process and stressing the need for careful decision making, sometimes resulting in contradictory practices, in relation to both short- and long-term goals.

Finally, the support team must also consider the instruction offered against the athlete's perceptions about what is happening during the execution. Typically, athletes do not understand their movement in quantitative terms (e.g., instructing the athlete to "increase/decrease your knee flexion by 10%"; Giblin *et al.*, 2015), but rather through sensory representation (e.g., feeling the body lean, sound of the foot making contact with the ball, etc.) and the perceived effort required. Indeed, understanding these cues from the athlete and feeding them back (e.g., "now do that again but ramp up the 'volume' on it"), a process akin to imagery response training (Lang *et al.*, 1980), will require consistency across the support team's language and/or only a limited number of individuals offering instruction. Notably, however, these cues may offer quality guidance in locating the general area of focus and then serve to reduce the number of tracking target variables down to a single kinematic measure; as opposed to tracking all components within the kinematic chain. Assessing that the contrast versions are at least being executed in the correct direction is a positive sign, one that should be much welcomed by both athlete and coach!

3.3 Stage 3: adjustment

Despite regaining high consciousness, success at achieving the desired kinematics is often infrequent, with the athlete normally generating an approximation (cf. Tallet *et al.*, 2008) of the desired target variable. Accordingly, the Adjustment stage attempts to gradually increase the accuracy of executions and heighten the athlete's acceptance and comfort towards the new technique. Carson, Collins and Jones (2014) describe this process as "shaping" (p. 69), whereby the motoric representation, in the form of kinaesthetic, visual and verbal stimuli, undergo progressive revisions/updates as the athlete becomes more familiar with, and better at, the targeted movement goal; that is, going beyond the initial sense of "this feels strange". Indeed, demonstrating technical improvements can be very motivational and provide an increasingly vivid perception of 'what' and 'how' to execute. As such, the biomechanist can play a crucial role here in assessing for any changes to drive the modification intervention.

As an exemplar practice of the shaping intervention, Carson, Collins and Jones (2014) employed self-modelling, but only against the athlete's best attempt. In simple terms, as the athlete got closer to the targeted technique, video footage was replaced in order to stimulate an ever-improving mental imagery prime (Holmes & Collins, 2001; Lang, 1979). It was crucial that the athlete did not get stuck part way through the change and automate an incomplete version or regress back toward the original version. In doing so, we highlight two important factors: (1) the viewing angle and (2) the level of mental engagement during observation. To maximise an observational effect, the athlete must be able to see their progress being made, after all, the behaviour is the intention of future attempts. As such, the modelling video might not necessarily be the same as used in conventional technique analysis (e.g., sagittal or frontal plane). Also, these images must relate to what the movement felt like during execution. Watching and recalling the cues attended to increases the vividness of the skill; therefore providing a greater number of retrieval cues for subsequent attempts (Paivio, 1986).

At the same time, the support team must intervene to ensure that the athlete departs from their previously erroneous technique; otherwise the risk of regression becomes increasingly likely. We recommend a tapering strategy for both physical and mental practice. As utilised within javelin throwing (Collins *et al.*, 1999), weightlifting (Carson, Collins & Jones, 2014) and golf (Carson & Collins, 2015), this requires the gradual removal of incorrect attempts within the contrast training regime, therefore increasing the pressure to execute with, and establishment of, the desirable kinematics.

3.4 Stage 4: (re)automation

Once the athlete can consistently achieve the desired new technique, the skill must return to being executed under largely subconscious control. Indeed, MacPherson *et al.* (2009) explain the effects of focussing on part-skill 'cues' as detrimental under conditions of competitive pressure due to the movement being fragmented, which disrupts the necessary flow and timing of the entire movement. Therefore, it is suggested that athletes focus on holistic patterns of thought which emphasise the whole action and do not overly tax attentional recourses; meaning that athletes can still utilise task-relevant environmental information (e.g., assessing the strength and direction of wind). Often these cognitions relate to the timing, or rhythm, of the movement and intensity/emotion contained within it (cf. Holmes & Collins, 2001). For instance, Collins *et al.* (1999) overlaid a sequence of bleeps onto real-time video footage for a javelin thrower when reautomating their technique. Bleeps occurred at every foot–ground contact; the volume, pitch and timing emphasised to reflect changes in the technical phases (i.e., straight run, sideways turn, planting the 'block' leg and throw) and how the athlete perceived these to be represented. Importantly, the stimulus was a fluid and continuous stream of information. Clearly there are similarities here to the skill of penalty kicking, where rhythm and timing appear to be distinctly emphasised (at least behaviourally) within the professional game (e.g., Neil Jenkins; see Jackson & Baker, 2001). Likewise, mood words (e.g., thump, swish, clip) can also provide the athlete with a beneficial motoric 'aide memoire' of the skill (Rushall, 1979), as long as it reflects pertinent movement capacities such as the required strength, speed, power, agility, balance or endurance. In either case, such thoughts are described as "sources of information" (MacPherson *et al.*, 2008, p. 289), providing a "prophylactic against potentially disruptive cognitions and emotional states that inhibit fluid movement" (MacPherson, *et al.*, 2009, p. S58) and as creating a most direct route to retrieval of the entire skill from memory (see Winter *et al.*, 2014). As such, and in contrast to experimentally-derived guidance to avoid thinking about one's body movements (Masters, 1992; Wulf, 2013), we encourage practitioners to consider the role that a positive self-focus (see Carson & Collins, 2016) can have in promoting better performance outcomes (Bortoli *et al.*, 2012).

Notably, regaining automaticity should be gradual, in contrast to the more catastrophic nature of the Awareness stage. Primarily, this is because technical components that were not targeted for refinement must 'settle' in with this new version. In preventing too quick a return to automaticity, the support team should taper out the intensity and frequency of their input by adopting a more hands-off approach within more representative, on-field and game-like contexts. That is not to say the movement will be fully established and always consistent from day-to-day, there will be some inevitable bumpiness, but that constant harassment from the biomechanist and/or coach regarding technical instruction will not help the matter. In fact, consultation and evaluation of skill progression would most probably be well suited to the sport psychologist's less threatening and emotionally-attuned role.

3.5 Stage 5: assurance

Competition at any level brings with it an expected degree of anxiety and, therefore, potential to influence an athlete's execution. Within the football context, anxiety is likely to manifest both physiologically (e.g., breathlessness, high heart rate, muscular tension and fatigue) and psychologically (e.g., worry, self-consciousness, negative self-focus). Indeed, anxiety has long been understood to be an essential component of optimum performance (Yerkes & Dodson, 1908), but also a significant debilitating source which can cause regression in motor control (e.g., Collins *et al.*, 2001; Pijpers *et al.*, 2003). For elite-level or professional athletes, the consequences of succumbing to these latter effects are often far more severe than within recreational or amateur contexts (i.e., potential to be dropped by the manager, social ridicule and personal embarrassment). Accordingly, the support team must proactively work to ensure that the athlete not only possesses a high degree of automaticity, but also confidence in this process (Carson & Collins, 2016). When these factors are high, the skill can almost certainly be pressure proofed against negative anxiety effects (Cheng *et al.*, 2009), leading to improved performance consistency and proficiency. A significant challenge confronting the athlete is, therefore, being able to achieve complete and fluid activation of the newly refined skill when distracted from multiple streams. While it is often recommended that the athlete try to clear their mind and think of nothing, this reality is normally impossible in these situations and akin to trying to consciously make oneself fall asleep (Montero, 2015). Instead, confidence in knowing how to 'make it happen' and, that one can make it happen, offers a much more realistic solution to preventing such an uncomfortable scenario occurring in the first place. Of course, this search for greater assurance in performance does not develop without practice.

From a practical perspective, a method that we find to be successful in securing skills at this stage, is to simulate anxiety and face the symptoms head on, alongside the provision of quality evaluation of both performance outcome and process consistency. We call this intervention combination training; that is, combining physical fatigue symptoms with a difficult level of technical challenge. Exemplar implementation has included sprints prior to executing challenging shot strategy in golf (Carson & Collins, 2015) and fully committed javelin throws (Collins *et al.*, 1999), introducing higher social pressure by the presence of peers (Carson, Collins & Jones, 2014) and, in our professional consultancy experience within rugby union, incorporating upper body weight lifting exercises alongside 150 m sprints prior to each line-out throw. Within the kicking context, manipulation of task difficulty may utilise acute kicking angles, longer distances and perceived social pressure from the presence of coaching and/or managerial staff (or at least video recording that the player believes will be shown to these individuals). In fact, it is not uncommon in our experience that, when athletes have established their skills to such a high degree, they perform better during this type of training when compared to unpressurised conditions. Crucially, an ability to offer objective – either three-dimensional and/or video – feedback to demonstrate the skill's security is a very powerful tool to assure the athlete just how consistent their performances really are. Additionally, verification of this kind is of equal importance in preventing the coach from implementing further intervention.

4. Assessing and tracking athlete performance

In light of our previous assertion that there is a lot to manage during the refinement process, biomechanists are well equipped to bring several valuable pieces of information to assist in tracking progress through the Five-A Model. Indeed, demonstration of accurate and meaningful data can present essential monitoring for other support disciplines and drive the necessary stages involved. Since many chapters within this book provide ample guidance towards the measurement of football movements, here we will introduce a related concept that has shown to bridge the cognitive–coordinative relationship explained by Schack and Bar-Eli (2007); inter-trial movement variability.

Firstly from a co-ordination perspective, a most fundamental investigation of this domain addresses how successful outcomes are consistently achieved by a redundant motor system (i.e., the degrees-of-freedom problem; Bernstein, 1967). In other words, how does the central nervous system (CNS) solve a movement problem, such as organising the limbs to kick a ball, when there are so many different whole body joint configurations available to it? Importantly in this regard, it is accepted that no two movements are ever executed in exactly the same way, even at the elite level where athletes and coaches train to ensure a high degree of establishment (e.g., Carson & Collins, 2016; MacPherson *et al.*, 2008). During skill acquisition, however, inter-trial movement variability can be seen to reduce as movements become both more efficient and proficient, due at least to reductions in stochastic noise (Bobrownicki *et al.*, 2015; MacPherson *et al.*, 2008; Müller & Sternad, 2004). For closed and self-paced skills, individually preferred movement patterns are stabilised with practice to exploit each individual's physical characteristics (hence why every kicker will have their own recognisable 'style') to complete the task requirements. Therefore, such variability can be considered as 'functional' (Davids *et al.*, 2003) and catering for the inevitably different task requirements such as kicking from different distances, angles and ground conditions.

How this variability is structured across all of the different movement components is more complex and dependent on the motor system apparatus (e.g., limb length and joint flexibility) involved. Recent interpretations have viewed motor redundancy not as problematic to the CNS (Bernstein, 1967), but instead as a luxury (Gelfand & Latash, 1998). According to the UnControlled Manifold (UCM) concept (Scholz & Schöner, 1999; Schöner, 1995), the CNS preferentially stabilises (i.e., reduces the variability) aspects of the movement that are essential to task success and frees up (i.e., increases the variability) less essential movement components to accommodate/support changes imposed by dynamic task constraints (e.g., kicking on wet vs. dry grass). The UCM concept therefore satisfies concerns that the CNS cannot control every movement component, and that it is an adaptable system within a dynamic environment. Crucially for sports biomechanists, this perspective carries with it a number of implications. Firstly, variability is not simply 'noise' within the system that should be ignored. Secondly, movement invariance does not reflect representative executions that should be sought after; in fact, too low variation could be a hallmark of dysfunctional movement control. Thirdly, and finally, the variability of specific movement components (e.g., pelvis–torso lateral flexion at foot–ball contact) may not be comparable between individuals.

Now we turn to the cognitive element of this tracking tool. Recently, we suggested that the co-variation principle explained by the UCM concept might apply also when movement components are subjected to different requirements of conscious intervention (Carson, Collins & Richards, 2014). Specifically, when an athlete decides to consciously emphasise the control of a movement component, they assign greater importance to it and, therefore, inter-trial variability would predictably decrease below that of normal functional levels. Concurrently, less associated aspects of that component would predictably increase in inter-trial variability due to a reduction in emphasis. Overall, this results in an imbalance of control across the entire skill; that is, 'dysfunctional' movement variability and a dip in performance (Carson & Collins, 2016). It is important to realise at this stage that the extent of this disparity cannot be quantitatively known in advance, only that measurement will never reach zero. We suggest that a plateau across several sessions following a noticeable decrease should be aimed for. Taking the skill refinement process in its entirety, therefore, initial inter-trial variability across different movement components would be predictably different (cf. Scholz & Schöner, 1999) but relatively consistent from session to session (i.e., a well-established movement pattern; Carson & Collins, 2016). Once applying a narrow internal focus of attention, overall control will be unbalanced (with the target variable reducing and less associated components increasing in inter-trial variability), until a time when the technique is modified and conscious attention is applied more holistically. Indeed, this disruption to the overall movement control once again highlights a risk involved and therefore need for careful planning to decide when the right time is to start refining. If all is going to plan however, pre-change variability levels offer a valuable reference guide to know when the athlete no longer needs to attend to the target variable and the new version has been internalised. Here, variability levels should return to normal functional amounts, with the new kinematics of course! Crucially at this stage, functional movement variability must also be demonstrated under pressure testing conditions for the refinement to be considered complete. As such, it is most beneficial for biomechanical instrumentation to be well suited to applied testing conditions, therefore offering a desirable alternative to self-reported measures of conscious control during the latter stages.

To exemplify this application, we have explored the practical utility of movement variability in the comparable closed and self-paced skill of golf. Specifically, our most recent research has assessed co-variation trends across several training designs and both short- and long-term timescales, with promising effect. For example, Carson, Collins and Richards (2014) showed high-level golfers to demonstrate greater consistency for target variables when intentionally executing non-preferred shot trajectories versus a more natural, less effortful and preferred type (fades or draws; i.e., left-to-right or right-to-left ball flights), while variability for contralateral non-target variables increased (see MacPherson *et al.*, 2008, for similar effects when employing a part-skill vs. holistic focus in javelin throwing). In another study evaluating the efficacy of different training environments when initiating refinement in the Awareness stage, variability of target variables reduced more when golfers executed shots in front of a net versus on a driving range with 100% outcome feedback (Carson *et al.*, 2016). These data suggested that better use of attentional control towards narrow internal cues was apparent in the absence of environmental distractions. Finally, Carson and Collins (2015) report longitudinal case studies showing different outcomes for high-level golfers attempting permanent and pressure-resistant refinements. Notably, the level of agreement with expected co-variation trends corresponded to the extent of inter-vention success. One participant was able to complete their intended refinement, and co-variation trends were largely as we have discussed. For another partici-pant, the kinematics were modified as planned, however there was a 'double-dip' in the target variable's variability signalling a reduction in conscious attention following the change and then a return to increased attention shortly after. Self-report data indicated that they were not yet fully comfortable with the new move-ment, probably due in part to intervention adherence problems. In a final example, the refinement was abandoned with the golfer not able to complete the change due to confusion, a lack of intention and experience of using the mental skills required. Movement variability trends in this case showed no resemblance to that expected. Therefore, employing movement variability has the potential to inform about possible derailments as well as intended progress. Crucially, however, interpreta-tion of data must consider biopsychosocial interactions to explain why training is/ is not working. As such, we believe that there warrants much anticipation towards what this 'psychomechanical' measure may offer football practitioners when designing and monitoring effective interventions.

5. Summary

The ability to successfully refine an already learnt and well-established skill is essential at times during high-performance sport. Indeed, success in this task requires necessary consideration of biopsychosocial factors which underpin development and, therefore, an interdisciplinary model that recognises the unique and interactive contributions of different specialists at varying points in the process. This chapter has explained how the biomechanist can usefully support and act within such a team to bring about effective refinement of closed motor skills, such as the rugby penalty kick. Central to the successful operation of an interdisciplinary team is the generation of an SMM derived from a PJDM approach. Facilitated by an atmosphere of open and intensive debate, the SMM ensures role clarity and an effective team dynamic within the supporting person-nel, while a clear and assured front is presented to the athlete–client. This chapter has also presented an overview of the Five-A Model (Carson & Collins, 2011) designed to facilitate the dual outcomes of long-term permanent and pressure proofed refinement. Beyond the traditional biomechanical focus on observation and diagnosis of errors, the Five-A Model first emphasises the need and methods to establish both cross-team consensus and athlete 'buy-in' regarding whether a refinement should be attempted, what needs refining and, if the decision to proceed is reached, how and when to proceed. Subsequent stages present a rationale and methodology for returning the movement to conscious control, shaping the movement towards the desired pattern, automating the modified pattern and assuring the athlete and coach that the refinement has been successfully accomplished. As such, the Five-A Model may be considered as an integrated, practical framework to guide the performer, coach and support team. For the biomechanist in particular, the model aids in understanding the objectives and activities of other support team members, and raises important considerations regarding what, how and when measurements are appropriate. Finally, this chapter has explained how the biomechanist can employ inter-trial variability within several different training environments and simulations to evaluate athlete progress throughout the nonlinear refinement process. Indeed, the utilisation of this measure can facilitate a better understanding of the cognitive–coordinative relationship and, therefore, provide valuable data for both sport psychologist and coach in relation to the athlete's attentional focus and automaticity. In conclusion, we hope that this chapter has stimulated discussion and offered new suggestions on how a biomechanist can act most efficiently within an interdisciplinary team when implementing technical refinement.

References

Abraham, A., & Collins, D. (2011). Taking the next step: Ways forward for coaching science. *Quest*, 63, 366–384. doi:10.1080/00336297.2011.10483687

Ajzen, I. (1991). The theory of planned behavior. *Organizational Behavior and Human Decision Processes*, 50, 179–211. doi:10.1016/0749-5978(91)90020-T

Bailey, R., Collins, D., Ford, P., MacNamara, Á., Toms, M., & Pearce, G. (2010). Participant development in sport: An academic review. Retrieved from Sports Coach UK: http://www.sportscoachuk.org/resource/participant-development-sport-academic-review

Beilock, S. L., Carr, T. H., MacMahon, C., & Starkes, J. L. (2002). When paying attention becomes counterproductive: Impact of divided versus skill-focused attention on novice and experienced performance of sensorimotor skills. *Journal of Experimental Psychology: Applied*, 8, 6–16. doi:10.1037/1076-898x.8.1.6

Bernstein, N. A. (1967). *The coordination and regulation of movements*. Oxford: Pergamon Press.

Bobrownicki, R., MacPherson, A. C., Coleman, S. G. S., Collins, D., & Sproule, J. (2015). Re-examining the effects of verbal instructional type on early stage motor learning. *Human Movement Science*, 44, 168–181. doi:10.1016/j.humov.2015.08.023

Bortoli, L., Bertollo, M., Hanin, Y., & Robazza, C. (2012). Striving for excellence: A multi-action plan intervention model for shooters. *Psychology of Sport and Exercise*, 13, 693–701. doi:10.1016/j.psychsport.2012.04.006

Burke, V. (2011). Organizing for excellence. In D. Collins, A. Button, & H. Richards (Eds.), *Performance psychology: A practitioner's guide* (pp. 99–119). Oxford: Elsevier.

Burwitz, L., Moore, P. M., & Wilkinson, D. M. (1994). Future directions for performance-related sports science research: An interdisciplinary approach. *Journal of Sports Sciences*, 12, 93–109. doi:10.1080/02640419408732159

Carson, H. J., & Collins, D. (2011). Refining and regaining skills in fixation/diversification stage performers: The Five-A Model. *International Review of Sport and Exercise Psychology*, 4, 146–167. doi:10.1080/1750984x.2011.613682

Carson, H. J., & Collins, D. (2015). Tracking technical refinement in elite performers: The good, the better, and the ugly. *International Journal of Golf Science*, 4, 67–87. doi:10.1123/ijgs.2015-0003

Carson, H. J., & Collins, D. (2016). The fourth dimension: A motoric perspective on the anxiety–performance relationship. *International Review of Sport and Exercise Psychology*, 9, 1–21. doi:10.1080/1750984X.2015.1072231

Carson, H. J., & Collins, D. (2016). Implementing the Five-A Model of technical change: Key roles for the sport psychologist. *Journal of Applied Sport Psychology*, 28, 392–409. doi:10.1080/10413200.2016.1162224

Carson, H. J., Collins, D., & Jones, B. (2014). A case study of technical change and rehabilitation: Intervention design and interdisciplinary team interaction. *International Journal of Sport Psychology*, 45, 57–78. doi:10.7352/IJSP2014.45.057

Carson, H. J., Collins, D., & Richards, J. (2014). Intra-individual movement variability during skill transitions: A useful marker? *European Journal of Sport Science*, 14, 327–336. doi:10.1080/17461391.2013.814714

Carson, H. J., Collins, D., & Richards, J. (2016). Initiating technical refinements in high-level golfers: Evidence for contradictory procedures. *European Journal of Sport Science*, 16, 473–482. doi:10.1080/17461391.2015.1092586

Cheng, W. N. K., Hardy, L., & Markland, D. (2009). Toward a three-dimensional conceptualization of performance anxiety: Rationale and initial measurement development. *Psychology of Sport and Exercise*, 10, 271–278. doi:10.1016/j.psychsport.2008.08.001

Collins, D., Bailey, R., Ford, P. A., MacNamara, Á., Toms, M., & Pearce, G. (2012). Three Worlds: New directions in participant development in sport and physical activity. *Sport, Education and Society*, 17, 225–243. doi:10.1080/13573322.2011.607951

Collins, D., Burke, V., Martindale, A., & Cruickshank, A. (2015). The illusion of competency versus the desirability of expertise: Seeking a common standard for support professions in sport. *Sports Medicine*, 45, 1–7. doi:10.1007/s40279-014-0251-1

Collins, D., Carson, H. J., & Cruickshank, A. (2015). Blaming Bill Gates AGAIN! Misuse, overuse and misunderstanding of performance data in sport. *Sport, Education and Society*, 20, 1088–1099. doi:10.1080/13573322.2015.1053803

Collins, D., & Collins, J. (2011). Putting them together: Skill packages to optimize team/group performance. In D. Collins, A. Button, & H. Richards (Eds.), *Performance psychology: A practitioner's guide* (pp. 361–380). Oxford: Elsevier.

Collins, D., & Hill, A. (2016). Shared mental models in sport and refereeing. In S. D. Obhi and E. S. Cross (Eds.), *Shared representations: Sensorimotor foundations of social life* (pp. 588–602). Cambridge: Cambridge University Press.

Collins, D., Jones, B., Fairweather, M., Doolan, S., & Priestley, N. (2001). Examining anxiety associated changes in movement patterns. *International Journal of Sport Psychology*, 32, 223–242.

Collins, D., Morriss, C., & Trower, J. (1999). Getting it back: A case study of skill recovery in an elite athlete. *The Sport Psychologist*, 13, 288–298.

Collins, D. J., Trower, J., & Randall, G. (2002, November). Preparing to win. Paper presented at the UKSI World Class Coaching Conference, The Belfry.

Davids, K., Glazier, P., Araújo, D., & Bartlett, R. (2003). Movement systems as dynamical systems: The functional role of variability and its implications for sports medicine. *Sports Medicine, 33,* 245–260. doi:10.2165/00007256-200333040-00001

Fitts, P. M., & Posner, M. I. (1967). *Human performance.* California: Brooks/Cole Publishing Company.

Gelfand, I. M., & Latash, M. L. (1998). On the problem of adequate language in motor control. *Motor Control, 2,* 306–313.

Giblin, G., Farrow, D., Reid, M., Ball, K., & Abernethy, B. (2015). Exploring the kinaesthetic sensitivity of skilled performers for implementing movement instructions. *Human Movement Science, 41,* 76–91. doi:10.1016/j.humov.2015.02.006

Godbout, A., & Boyd, J. E. (2010). Corrective sonic feedback for speed skating: A case study. Paper presented at the 16th International Conference on Auditory Display, Washington. Retrieved from https://smartech.gatech.edu/handle/1853/49865

Hanin, Y., Korjus, T., Jouste, P., & Baxter, P. (2002). Rapid technique correction using old way/new way: Two case studies with Olympic athletes. *The Sport Psychologist, 16,* 79–99.

Hanin, Y., Malvela, M., & Hanina, M. (2004). Rapid correction of start technique in an Olympic-level swimmer: A case study using old way/new way. *Journal of Swimming Research, 16,* 11–17.

Hill, D. M., & Shaw, G. (2013). A qualitative examination of choking under pressure in team sport. *Psychology of Sport and Exercise, 14,* 103–110. doi:10.1016/j.psychsport.2012.07.008

Holmes, P. S., & Collins, D. J. (2001). The PETTLEP approach to motor imagery: A functional equivalence model for sport psychologists. *Journal of Applied Sport Psychology, 13,* 60–83. doi:10.1080/10413200109339004

Jackson, R. C., & Baker, J. S. (2001). Routines, rituals, and rugby: Case study of a world class goal kicker. *The Sport Psychologist, 15,* 48–65.

Lang, P. J. (1979). A bio-informational theory of emotional imagery. *Psychophysiology, 16,* 495–512. doi:10.1111/j.1469-8986.1979.tb01511.x

Lang, P. J., Kozak, M. J., Miller, G. A., Levin, D. N., & McLean Jr, A. (1980). Emotional imagery: Conceptual structure and pattern of somato-visceral response. *Psychophysiology, 17,* 179–192. doi:10.1111/j.1469-8986.1980.tb00133.x

MacNamara, Á., Button, A., & Collins, D. (2010). The role of psychological characteristics in facilitating the pathway to elite performance Part 1: Identifying mental skills and behaviors. *The Sport Psychologist, 24,* 52–73.

MacNamara, Á., Holmes, P., & Collins, D. (2008). Negotiating transitions in musical development: The role of psychological characteristics of developing excellence. *Psychology of Music, 36,* 335–352. doi:10.1177/0305735607086041

MacPherson, A. C., Collins, D., & Morriss, C. (2008). Is what you think what you get? Optimizing mental focus for technical performance. *The Sport Psychologist, 22,* 288–303.

MacPherson, A. C., Collins, D., & Obhi, S. S. (2009). The importance of temporal structure and rhythm for the optimum performance of motor skills: A new focus for practitioners of sport psychology. *Journal of Applied Sport Psychology, 21,* 48–61. doi:10.1080/10413200802595930

Martindale, A., & Collins, D. (2005). Professional judgment and decision making: The role of intention for impact. *The Sport Psychologist, 19,* 303–317.

Masters, R. S. W. (1992). Knowledge, knerves and know-how: The role of explicit versus implicit knowledge in the breakdown of a complex motor skill under pressure. *British Journal of Psychology, 83,* 343–358. doi:10.1111/j.2044-8295.1992.tb02446.x

Montero, B. G. (2015). Is monitoring one's actions causally relevant to choking under pressure? *Phenomenology and the Cognitive Sciences*, 14, 379–395. doi:10.1007/s11097-014-9400-0

Müller, H., & Sternad, D. (2004). Decomposition of variability in the execution of goal-oriented tasks: Three components of skill improvement. *Journal of Experimental Psychology: Human Perception and Performance*, 30, 212–233. doi:10.1037/0096-1523.30.1.212

Oudejans, R. R. D., Koedijker, J. M., & Beek, P. J. (2007). An outside view on Wulf's external focus: Three recommendations. *E-journal Bewegung und Training*, 1, 41–42. www.ejournal-but.de Retrieved from www.ejournal-but.de

Paivio, A. (1986). *Mental representations: A dual-coding approach*. New York: Oxford University Press.

Pijpers, J. R., Oudejans, R. R. D., Holsheimer, F., & Bakker, F. C. (2003). Anxiety–performance relationships in climbing: A process-oriented approach. *Psychology of Sport and Exercise*, 4, 283–304.

Pinder, R. A., Davids, K., Renshaw, I., & Araújo, D. (2011). Representative learning design and functionality of research and practice in sport. *Journal of Sport and Exercise Psychology*, 33, 146–155.

Rushall, B. S. (1979). *Psyching in sports*. London: Pelham Books.

Schachter, S. (1982). Recidivism and self-cure of smoking and obesity. *American Psychologist*, 37, 436–444.

Schack, T., & Bar-Eli, M. (2007). Psychological factors of technical preparation. In B. Blumenstein, R. Lidor, & G. Tenenbaum (Eds.), *Psychology of sport training* (pp. 62–103). Münster, Germany: Meyer & Meyer Sport.

Scholz, J. P., & Schöner, G. (1999). The uncontrolled manifold concept: Identifying control variables for a functional task. *Experimental Brain Research*, 126, 289–306. doi:10.1007/s002210050738

Schöner, G. (1995). Recent developments and problems in human movement science and their conceptual implications. *Ecological Psychology*, 7, 291–314. doi:10.1207/s15326969eco0704_5

Winter, S., MacPherson, A. C., & Collins, D. (2014). "To think, or not to think, that is the question". *Sport, Exercise, and Performance Psychology*, 3, 102–115. doi:10.1037/spy0000007

Wulf, G. (2013). Attentional focus and motor learning: A review of 15 years. *International Review of Sport and Exercise Psychology*, 6, 77–104. doi:10.1080/1750984x.2012.723728

Yerkes, R. M., & Dodson, J. D. (1908). The relation of strength of stimulus to rapidity of habit-formation. *Journal of Comparative Neurology and Psychology*, 18, 459–482.

Part V
Artificial turf

16 Critical issues of shock absorbing property and its detection test of long pile artificial turf in football

Hiroyuki Nunome

1. Introduction

The use of long pile third generation artificial turf (3-g turf) is becoming very popular for sports that are usually played on natural grass, particularly soccer and rugby. The Fédération Internationale de Football Association (FIFA) and the International Rugby Board (IRB) recently permitted the use of 3-g turf for official and international tournaments. In soccer, all the matches in recent two female world championships (2014 FIFA U-20 Women's World Cup; 2015 FIFA Women's World Cup) were played on 3-g turf pitches.

In contrast to earlier artificial turf generations, it has been reputedly said that 3g artificial turfs possess similar mechanical properties with those of natural turf. A considerable number of cohort studies reported that there are no clear differences for injury risk (Ekstrand *et al.*, 2006; Fuller *et al.*, 2010; Fuller *et al.*, 2007; Steffen *et al.*, 2007) and players' movement pattern (Andersson, *et al.*, 2008) between newer 3-g artificial turf and natural grass surfaces. However, coaches and players still have concerns that some of the mechanical properties of 3-g artificial turf surfaces may be linked to acute or chronic sports injuries. In fact, 2015 FIFA Women's World Cup adopted 3-g turf as their exclusive playing surface, there was some dispute about its playability and safety among coaches, players and sports scientists.

What are pertinent properties required for sports surfaces? Relevant sport surface properties should include adequate cushioning ability, friction character- istics and influence on energy loss (Dixon *et al.*, 1999). Of these aspects, as it is often assumed that specific sports injuries are a direct consequence of surfaces that are very hard (Nigg, 1990; Williams, Hume & Kara, 2011), the cushioning ability is considered as the most important property for sports surfaces.

In this chapter, therefore, we focus on shock absorbency of 3-g turfs to deepen our understanding of: 1) how this property has been tested and approved by sports governing bodies and 2) how close 3-g turf replicates the property of natural grass in the playing field.

2. Standard test

Artificial Athlete Berlin, described in the German Standard DIN 18032 (DIN test), has been the most widely used test to measure the shock absorbency of sports surfaces such as basketball playing floors (International Basketball Federation, 2010) and running tracks (International Association of Athletics Federations, 2009). FIFA and IRB have adopted the DIN test for the quantification of 3-g turf soccer and rugby fields (FIFA, 2008; IRB, 2008), however, in fact, little is known about how this test is suited to measuring the shock absorbency of 3-g turfs.

As shown in Figure 16.1 (top panel), in this test, a 20 kg mass is released from 55 mm height onto a spring and the force applied to the samples below is recorded by a load cell installed to the test foot. To date, several studies (Blackburn *et al.*, 2005; Dixon *et al.*, 1999; Nigg, 1990) already suggested that this conventional testing procedure does not reflect actual loading actions occurring in sports movements and has considerable technical disadvantages: 1) of not taking into account possible inertia terms of the test foot on surface samples, 2) of not providing any information about time-force profile during loading. As the validity of this standard test was very questionable, we made an attempt to re-examine how valid this conventional mechanical testing procedure is to evaluate the shock absorbency of 3-g turf (Nunome *et al.*, 2014).

The DIN test was conducted for all 3-g turf samples with different infill components: materials, the number of layers and rubber grain sizes (Table 16.1). The test was repeated five times at different un-touched, fresh spots on the surface of 3-g turf samples. The impact force was sampled at 10 kHz using an analog data acquisition conversion system (PowerLab, ADInstruments Ltd, New Zealand). The onset of impact loading was determined when the loading rate distinctively exceeded (more than twice) the range of noisy background oscillations. Conventionally, the data would be smoothed using a low-pass filter, however, to observe the raw loading changes in detail, no smoothing procedures were applied in the current test.

The average (±SD) impact loading changes of the DIN test are shown in Figure 16.2. In each panel, changes are shown against that of the most standard type of 3-g turf with four layers of sand and rubber infill (#1). Surprisingly, the DIN test produced bumpy, multiple peak force curves except for the 3-g turf with sand infill (#7) and failed to statistically distinguish (from the peak force) the shock absorbing property between 3-g turfs with different infill components other than sample #7.

It must be noted as a critical issue that the first impact peak occurred at the exact same timing, magnitude and loading rate over different 3-g turf samples. This indicates that the first impact peak does not represent any property of samples measured underneath but reflects some instrumental property inside the measuring system. After the first peak, the force curves suddenly declined to 0 values, suggesting that the load cell set on the test foot lost its contact with the dropped mass and oscillated largely between the dropped mass and the surface of the 3-g turfs. Finally, the standard DIN test was found to be, fundamentally, not appropriate for evaluating the shock absorbency of 3-g turfs due to its structural incompatibilities.

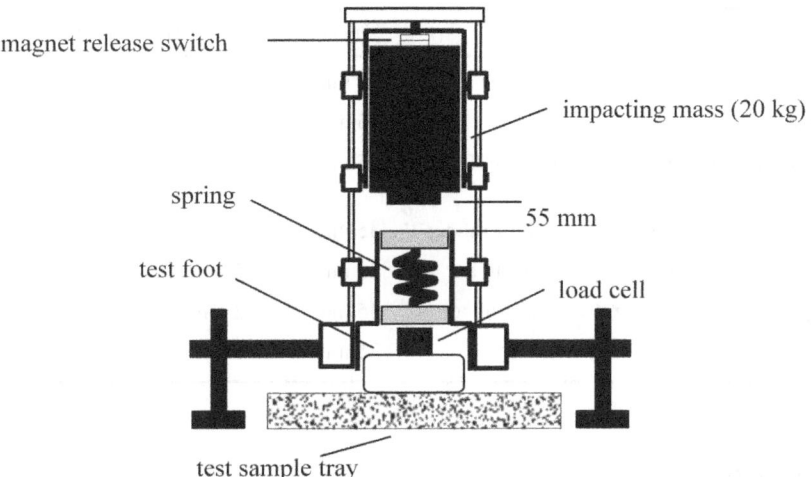

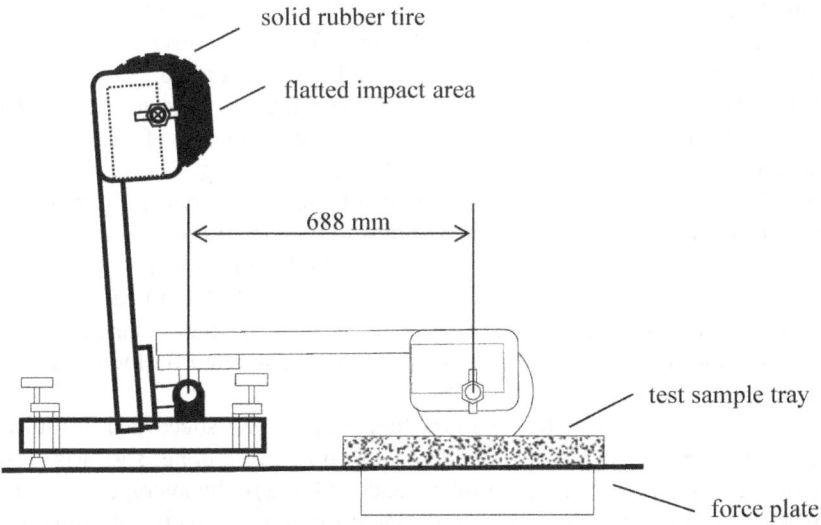

Figure 16.1 A simplified schema of experimental set up of the DIN test (top panel) and the new test (bottom panel) (reused with permission from Nunome, H., Inoue, K., Shinkai, H., Kozakai, R., Suito, H. and Ikegami, Y. "A novel comparison between standard and new testing procedures to assess shock absorbency of third generation artificial turfs" in *Sports Engineering*, published on 1st January, 2013, Springer)

Table 16.1 Details of infill component of tested 3-g turfs.

Groups	Infill component	Thickness	Length of artificial fiber
#1 4 layers	sand/rubber/sand/rubber	35 mm	62 mm
#2 2 layers	sand/rubber	35 mm	62 mm
#3 New type	urethane shock pad + TPE infill	35 mm	35 mm (planted on the shock pad)
#4 R rubber	regular grain (< 2 mm) rubber	35 mm	62 mm
#5 L rubber	large grain (< 3 mm) rubber	35 mm	62 mm
#6 S rubber	small grain (< 1 mm) rubber	35 mm	62 mm
#7 Sand	sand	35 mm	62 mm

3. New high loading test

By accepting the inappropriateness of the standard test, a new testing rig was developed to mimic a high impact load similar to human sporting actions (Figure 16.1, bottom panel). Hard landing from a representative height was selected because this action will cause the biggest impact on playing surfaces. Pilot data was collected to assess the impact load of hard human landing to obtain the baseline. The hard landing was performed by four healthy male subjects (height = 174.3 ± 6.3 cm; weight = 71.6 ± 3.5 kg) from a landing height of 55 cm, which simulates the height attained in a soccer jumping header (Wisløff *et al.*, 2004; Chamari *et al.*, 2004). That yielded one clear peak loading curve with magnitudes between 11.01 to 13.86 kN (15.6 to 19.2 times of body weight). Of the many physical parameters that could have been targeted, the new test was designed to reproduce the peak magnitudes during the severe human landing conditions. This new test was conducted on the same 3-g turf samples used for the standard test.

As shown in Figure 16.3, the new high loading test succeeded in illustrating clear differences for the shock attenuation property among 3-g turfs. In most cases (except for two layered rubber/sand infill (#2)), the average ± 1SD ranges were clearly apart from that of the standard type (#1). Moreover, although the initial loading curve of the new type (#3) closely matched with that of the standard type (#1), those two curves can be distinguished under high loading condition around the peak magnitudes. That result might validate the assertion that 3-g turf should be tested in more high loading conditions in terms of players' safety.

4. Natural turf properties

As mentioned earlier, the shock absorbency of natural turfs could have been an important target property for artificial turfs. However, little is known about its actual characteristics. To the best of our knowledge, there is only one study which

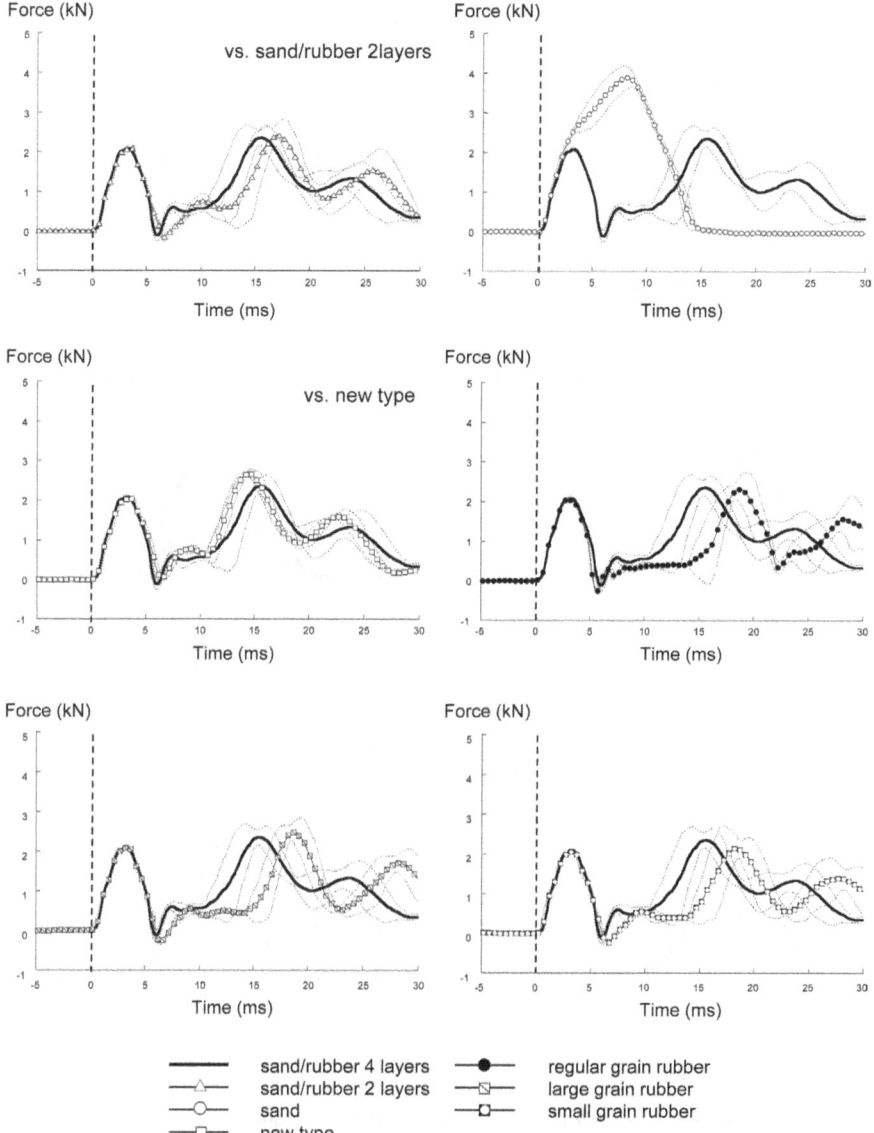

Figure 16.2 The average (±SD) impact loading changes of the DIN test. In each panel, changes are shown against that of the most standard type of 3-g turf (#1) (reused with permission from Nunome, H., Inoue, K., Shinkai, H., Kozakai, R., Suito, H. and Ikegami, Y. "A novel comparison between standard and new testing procedures to assess shock absorbency of third generation artificial turfs" in *Sports Engineering*, published on 1st January, 2013, Springer)

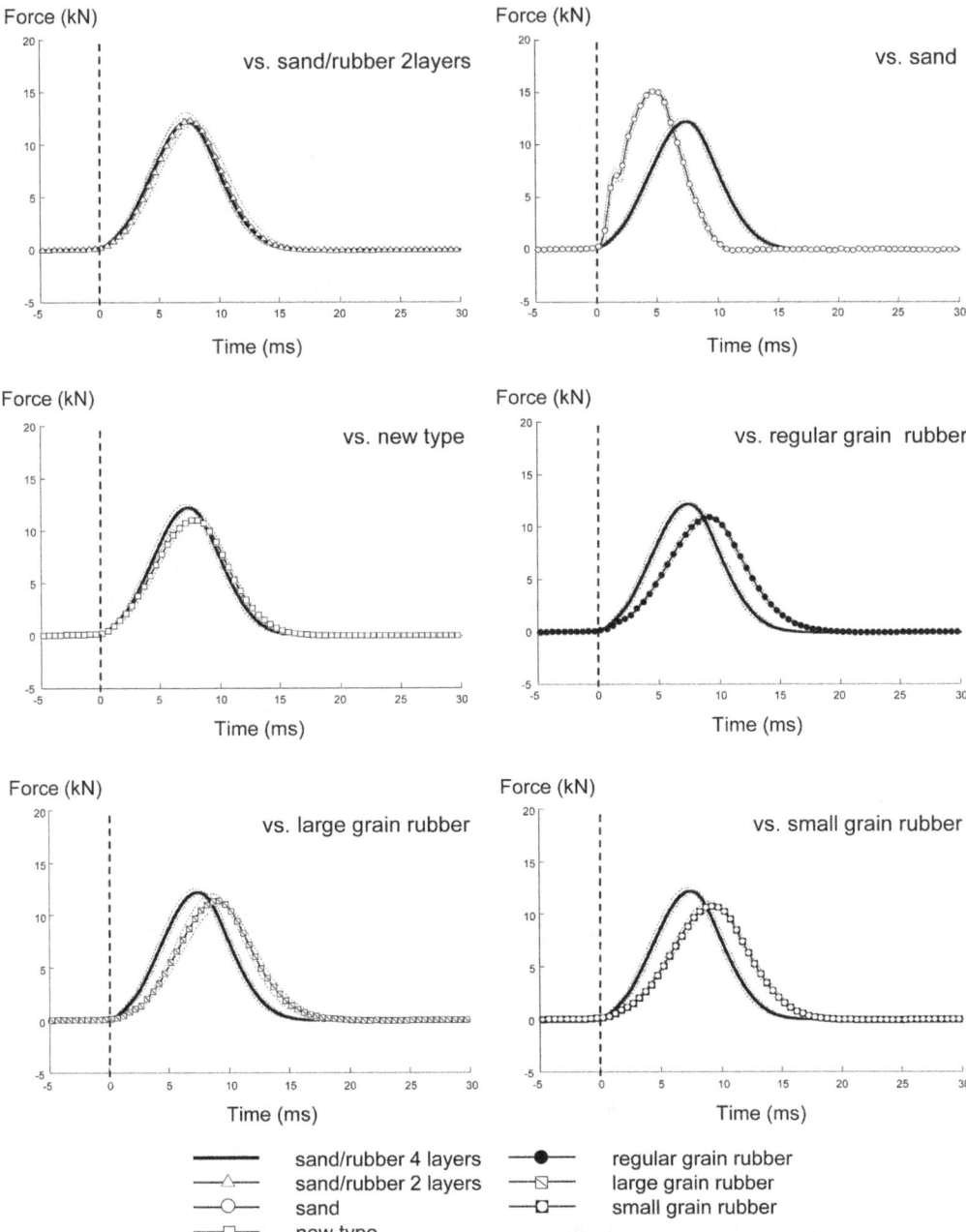

Figure 16.3 The average (±SD) impact loading changes of the new test. In each panel, changes are shown against that of the most standard type of 3-g turf (reproduced with permission from Nunome, H., Inoue, K., Shinkai, H., Kozakai, R., Suito, H. and Ikegami, Y. "A novel comparison between standard and new testing procedures to assess shock absorbency of third generation artificial turfs" in *Sports Engineering*, published on 1st January, 2013, Springer)

compared the property between natural turfs and 3-g turfs (Nunome, Inoue & Ikegami, 2014). To measure the natural turf property in outfield, we modified the new test rig (Figure 16.1, bottom panel) by installing an accelerometer and a rotary encoder to measure shock absorbency, loaded deformation from the surface of samples. As the linearity between the acceleration changes measured from the surface and that of loading force measured from underneath using a force platform has been verified ($R^2 > 0.990$), the modified version of the test allowed the measurement surface properties in outfield situations. It is evident that those profiles of natural turf would provide a significant insight for improving or fine-tuning the property of conventional 3-g artificial turfs.

We tested two types of natural turfs: transplanted with sod (nTRS) and over-seeded on sand base (nOVS), which were grown in specially made test field. Also, a natural turf regularly maintained and used for professional football league matches (nFTM) was tested. These properties were compared with those of 3-g turfs with sand-rubber mixed (aSRM), rubber (aRUB) and sand (aSAD) infills. These are corresponded to sample #1, #2 and #7 in Table 16.1. Also, to look at the effect of aging, an eight-year-old 3-g turf field (aOLD) infilled with mixed sand and rubber (#1) was additionally tested.

Figure 16.4 shows the average acceleration-deflection (stress-strain) curves of the natural and the 3-g artificial turfs. In each panel, changes are shown against that of the well maintained and regularly used natural turf in football stadium (nFTM) as an ideal target. In each curve, the top curve started from the origin show on-loading stress-strain aspect and the bottom curve demonstrate that of off-loading. A gap area between the two curves represents hysteresis, in other words, energy loss.

Overall, the 3-g turfs with rubber (aRUB) and rubber/sand mixed (aSRM) infills showed a similar peak loads to that of the well maintained natural turf (nFTM). However, the change of natural turfs (nFTM) indicated that substantial gaps still exist between natural and 3-g turfs for their stress-strain properties. These two types of 3-g turfs (aRUB and aSRM) showed more compliant features to the given load in most parts of the on-loading phase. Also, their surface deformations are restored to their original state during the off-loading phase while distinctive plastic surface deformations remained for the natural turf (nFTM). Moreover, the natural turf (nFTM) was characterized by distinctively larger hysteresis than these two artificial turfs. It can be suggested that conventional artificial turfs tend to return more energy to the soccer ball or a player's body during the off-loading phase while those energies are more readily absorbed in natural turfs. This property is most likely due to the elasticity of the rubber used as infills, which may account for higher ball rebounds typically seen on artificial turfs. On the other hand, a larger range of energy loss seen in the natural turfs might be more likely to limit a player's performance. Regarding this aspect, we are yet unable to explain why soccer players feel greater physical efforts on 3-g turfs (Anderson *et al.*, 2008). In general, softer surfaces, such as sand on the beach, demand more energy for most sporting actions (e.g. running, sprinting.) required for soccer. As Nigg (1990) demonstrated, point-elastic surfaces like 3-g turfs do not systematically react to

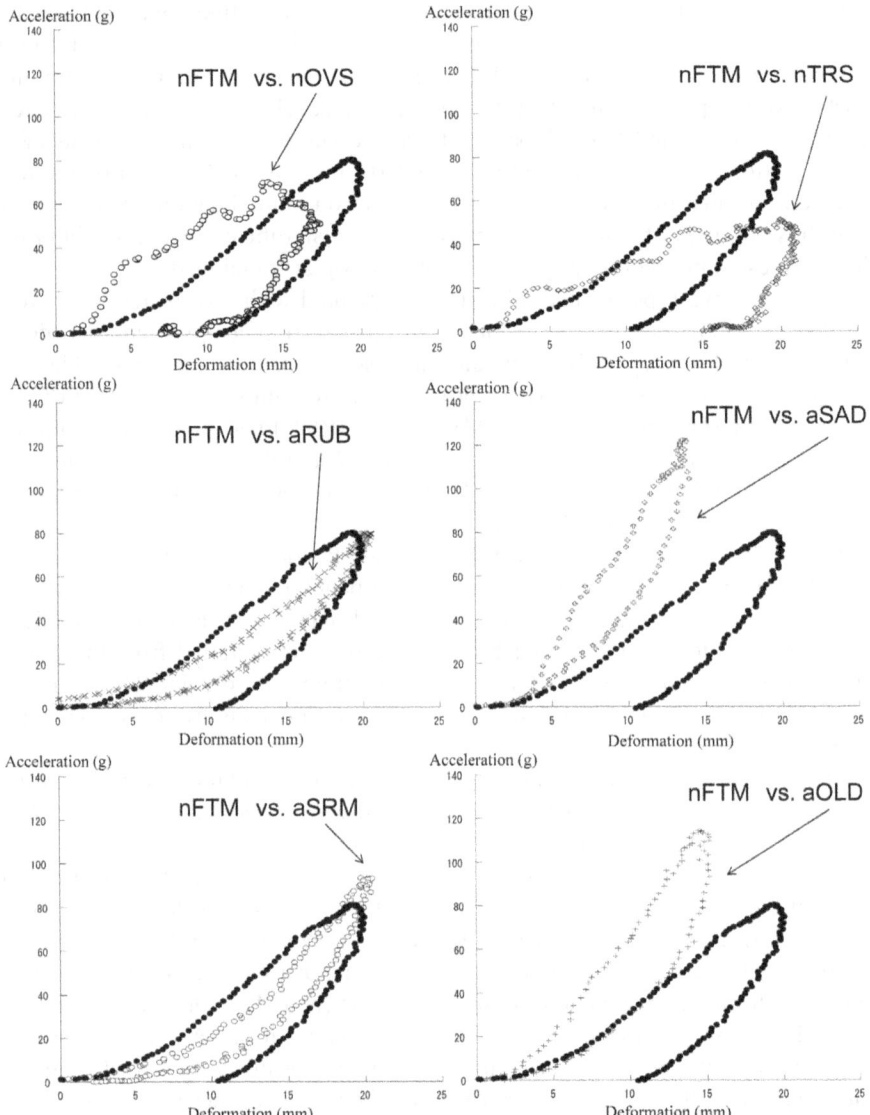

Figure 16.4 The average acceleration-deflection (stress-strain) curves of natural and 3-g artificial turfs. In each panel, changes are shown against that of the well maintained and regularly used natural turf in a football stadium

different loads. Thus, to clarify this dilemma, natural and 3-g turfs should be tested in a lower magnitude relevant for running or sprinting. Hence, further studies are still in need to examine the effect of playing surface on players' performance and perceptions using different test settings.

It is worth noting that the property of natural turfs can be changed drastically through daily maintenance. As shown in the two top panels in Figure 16.4, the properties of natural turfs are distinctively different. Obviously, nTRS has more compliant characteristics during on-loading and left a larger plastic deformation after off-loading than those of nFTM. In contrast, nOVS showed less compliant characteristics during on-loading, however, its peak load was more largely attenuated than that of nFTM. These results suggested that daily maintenance has a big impact on the stress-strain property of natural turfs which most likely changes the property of the bases (density of sod and fastened roots). This variable feature of natural turfs must be taken into account whenever we try to test or evaluate its shock absorbency.

It is known that aged 3-g turfs become much harder than their initial state due to compacting infills. However, to date, there has been no clear evidence to prove that change. The current result is the first quantitative data which clearly illustrates the deterioration of 3-g turf shock absorbency over time. As shown in the right bottom panel of Figure 16.4, the stress-strain properties of a 3-g turf frequently used for multiple sports activities for eight years was evaluated. The aged 3-g turf (aOLD) showed a remarkable discrepancy of the stress-strain personality from that of well-maintained natural turf (nFTM) and of fresh 3-g turf with exact same infill components (aSRM) (the left bottom panel). This result clearly indicated that aged 3-g turfs become distinctively harder than their initial state and tend to have a larger hysteresis. It can be assumed that repeated compression most likely deteriorated the elasticity of rubber infill grains. To suppress the deterioration observed for the most standard type of 3-g turf is a challenging assignment for 3-g turf producers to solve in the future.

5. Conclusion

A series of shock absorbency test clarified: 1) an inappropriateness of DIN 18032 to evaluate the shock absorbency of 3-g turfs due to its structural incompatibilities, 2) a high loading test that mimics human hard landing is useful to distinguish differences in the shock attenuation properties of third generation artificial turfs, 3) a considerable gap still exists for the stress-strain property between natural and third generation artificial turfs and 4) natural turf property can be drastically changed by daily maintenance.

References

Andersson, E., Ekblom, B. & Krustrup, P. (2008). Elite football on artificial turf versus natural grass: Movement patterns, technical standards, and player impressions. *Journal of Sports Sciences*, 26, 113–122.

Árnason, Á., Gudmunsson, Á., Dahl, H. A. & Johannsson, E. (1996). Soccer injuries in Iceland. *Scandinavian Journal of Medicine and Science in Sports*, 6, 40–45.

Blackburn, S., Nicol, A. C. & Walker, C. (2005). Development of a biomechanically validated turf testing rig. In: *Proceedings of 20th Congress of the International Society of Biomechanics* (p. 120). Ohio: Cleveland State University.

Chamari, Y., Hachana, Y., Ahmed, Y. B., Galy, O., Sghaïer, F., Chatard, J. C., Hue, O. & Wisløff, U. (2004). Filed and laboratory testing in young elite soccer players. *British Journal of Sports Medicine*, 38, 191–196.

DIN (1991). DIN Standard 18032-2 Part II: Sports Halls, Halls for gymnastics, games and multi-purpose use. Part 2: Sports floors, requirements and testing.

Dixon, S. J., Batt, M. E. & Collop, A. C. (1999). Artificial playing surfaces research: A review of medical, engineering and biomechanical aspects. *International Journal of Sports Medicine* 20, 209–218.

Durá, J. V., Hoyos, J. V., Lazano, L. & Martínez, A. (1999). The effect of shock absorbency sports surfaces in jumping. *Sports Engineering*, 2, 103–108.

Ekstrand, J., Timpka, T. & Hägglund, M. (2006). Risk of injury in elite football played on artificial turf versus natural grass: A prospective two-cohort study. *British Journal of Sports Medicine*, 40, 975–980.

Fédération Internationale de Football Association (2008). FIFA quality concept for football turf. http://www.fifa.com/mm/document/afdeveloping/pitch&equipment/50/15/94/footballturfbookletenglish_07012009.pdf. Accessed 31st August, 2012.

Fuller, C. W., Clarke, L. & Molloy, M. G. (2010). Risk of injury associated with rugby union played on artificial turf. *Journal of Sports Science*, 28, 563–570.

Fuller, C. W, Dick, R. W., Corlette, J. & Schmalz, R. (2007a). Comparison of the incidence, nature and cause of injuries sustained on grass and new generation artificial turf by male and female football players. Part 1: match injuries. *British Journal of Sports Medicine*, 41, i20–i26.

Fuller C. W., Dick R. W., Corlette, J. & Schmalz, R. (2007b). Comparison of the incidence, nature and cause of injuries sustained on grass and new generation artificial turf by male and female football players. Part 2: training injuries. *British Journal of Sports Medicine*, 41, i27–i31.

International Association of Athletics Federations (2009). IAAF Certification System: Track Facilities Testing Protocols. http://www.iaaf.org/mm/Document/ Competitions/TechnicalArea/04/85/77/20110414053903_httppostedfile_CertSys11_2_24133.pdf. Accessed 31st August, 2012.

International Basketball Federation (2010). Official Basketball Rules 2010. http://www.fiba.com/downloads/Rules/2010/OfficialBasketballRules2010.pdf. Accessed 31st August, 2012.

International Rugby Board (2008). Regulation 22: Standard relating to the use of artificial playing surfaces. http://www.irb.com/mm/document/lawsregs/0/regulation22080611_6080.pdf. Accessed 31st August, 2012.

Nigg, B. M. (1990). The validity and relevance of tests used for the assessment of sports surfaces. *Medicine and Science in Sports Exercise*, 22, 131–139.

Nunome, H., Inoue, K., Shinkai, H., Kozakai, R., Suito, H. & Ikegami, Y. (2014). A novel comparison between standard and new testing procedures to assess shock absorbency of third generation artificial turfs. *Sports Engineering*, 17,103–112.

Nunome, H., Inoue, K. & Ikegami, Y. (2014). High-loading absorption characteristics of natural and artificial turfs. In K. Sato, W. A. Sands & S. Mizuguchi (Eds.), *Proceedings of the 32nd International Conference of Biomechanics in Sports* (pp. 324–327).

Steffen, K., Andersen, T. E. & Bahr, R. (2007). Risk of injury on artificial turf and natural grass in young female football players. *British Journal of Sports Medicine*, 41, i33–i37.

Wisløff U., Castagna C., Helgerud J., Jones R., & Hoff, J. (2004) Strong correlation of maximal squat strength with sprint performance and vertical jump height in elite soccer players. *British Journal of Sports Medicine*, 38, 285–288.

17 Artificial turf in football

An injury perspective

Gerda Strutzenberger, Len Nokes and Gareth Irwin

1. Introduction

The global introduction of artificial playing surfaces in football has been promoted by the international governing body (Fédération Internationale de Football Association, FIFA) for over a decade as an alternative to natural grass surfaces and artificial surfaces are now used at the highest level of the game (e.g. Female Football World Cup 2015). Following definitions of the term injury, this chapter begins by exploring the epidemiological evidence surrounding injury occurrence in terms of its nature and location. Three main sections then follow; firstly examining musculoskeletal injury at specific sites (foot, ankle, knee, lower back and head); secondly, non-musculoskeletal injury including skin abrasion and infection, carcinogenic risk and airway irritation. The final section explores extrinsic factors that may influence injury risk on artificial turf including: shoe–surface interface, using both artificial and natural turf interchangeably, volume of exposure to artificial turf, climate, and football code. Finally, the chapter ends by drawing together some key observations, future directions and concluding remarks. Throughout the chapter the terms synthetic turf/surface, artificial turf/surface, football turf/surface are interchangeably used for artificial playing surfaces and will be referred to as artificial turf (AT) within this context.

2. Football injuries in general

Injuries can be differentiated as *traumatic injuries* (contact or non-contact), which are caused by "acute onset", and as *overuse injuries*, which are defined as "a pain syndromes of the musculoskeletal system with insidious onset and without any known trauma or disease that might have given previous symptoms" (Walden *et al.*, 2005, p. 6). The injury consensus statement for soccer promotes the injury rate to be expressed as injuries per 1000 hours of exposure (Fuller *et al.*, 2006). In football approximately 71% of all injuries are traumatic, (Ekstrand *et al.*, 2011a), while the number of overuse injuries varies between 9–40% (Mallo *et al.*, 2011, Dvorak & Junge, 2000, Nielsen & Yde, 1989, Ekstrand *et al.*, 2011a). Additionally, the definition of injury can be based on

1. *a time loss aspect*, considering "any injury occurring during scheduled training sessions or matches causing the player to miss the next training session or

match" (Walden et al., 2005, p. 119, based on Ekstrand, 1982). Additionally the severity of the injury can be classified as major (missing >28 days of training and match play), moderate (8–28 days) or minor (<3 days).

2. *a tissue injury aspect*, considering "any tissue damage caused by football regardless of subsequent absence from matches or training sessions" (Walden et al., 2005, p. 6).

3. *a need for medical assistance* aspect, considering injuries that require medical treatment (Hagglund et al., 2005).

Each of these approaches has advantages and disadvantages. For example, with the *"time loss"* and *"medical assistance"* definitions, minor injuries such as abrasion, muscle soreness or lower back pain, with which an athlete usually would not seek medical consolidation or sustain from training or match play, are not registered, and hence might give an uncomplete representation of possible effects of, for example, the surface on the well-being of the athletes.

In a 2013 EU report, football was ranked as main cause for team-sport-related injuries, compromising 71% of 2.5 million injuries occurring in this category (EuroSafe, 2013). In a regional example from Switzerland, Majewski *et al.* (2006) identified football being the main cause of 35% of all treated knee sport injuries in their hospital over a 10-year period. These numbers suggest that football is a dangerous sport activity. In fact though, these numbers are likely to have originated from the high population of people participating in this sport, which is with about 270 million players worldwide (FIFA, 2007) the most popular sport activity. The injury risk relative to the number of active member reveals a different picture. From this perspective, the injury ratio for football in the previous Swiss example is 0.32 (knee traumas/active members) and ranks football as 12th dangerous sport, being similar to handball, badminton, dance and volleyball (knee traumas/active members between 0.27–0.36) (Majewski *et al.*, 2006). From a perspective of injury rates per 1000 playing hours, the incidence of football injuries in adult male players is estimated by well-established epidemiologic studies to be between 6.2–35 per 1000 playing hours (e.g. Bengtsson *et al.*, 2013, Ekstrand *et al.*, 2011a, Dvorak and Junge, 2000).

The rate of injuries depends on several factors: age, level of competition, position on the field, environmental settings, location of injury, time in the season, gender (Silvers-Granelli *et al.*, 2015, Dvorak & Junge, 2000) and match vs training play with an increased risk at match play (e.g. Walden *et al.*, 2011b, Walden *et al.*, 2011a, Agel *et al.*, 2007, Dick *et al.*, 2007). Properties of natural turf surfaces that may be connected with injury prevalence include inappropriate friction characteristics, hardness and unevenness (Ekstrand & Nigg, 1989). The injuries most commonly reported involve the lower extremities and account for more than two thirds of men's (Agel *et al.*, 2007) and women's (Dick *et al.*, 2007) football injuries. The injuries typically consist of mild to moderate sprains, strains or contusions (Fuller *et al.*, 2006, Mallo *et al.*, 2011, Dvorak & Junge, 2000). Muscle injury constitutes approximately 31% (Ekstrand *et al.*, 2011b) of all injuries and occur at a rate of 92% (Ekstrand *et al.*, 2011b) in the four main groups of the lower limb: hamstrings, adductors, quadriceps and calf muscles

(Mallo *et al.*, 2011; Ekstrand *et al.*, 2011b). They cause between 27% (Ekstrand *et al.*, 2011b) to 43% (Mallo *et al.*, 2011) of total injury absence.

Knee and ankle joint injury cause approximately 30% of absence (Mallo *et al.*, 2011). Male players most commonly sustain an ankle ligament sprain (17%) during football practices and games, followed by an internal knee derangement (11%) during match play (Agel *et al.*, 2007), while female players most commonly sustain ankle ligament sprains (18%) and knee internal derangements (15.9%) during match play, while during practice upper leg muscle-tendon strains (21.3 %) and ankle ligament sprains (15.3 %) most commonly occur (Dick *et al.*, 2007). Head injuries have been shown to account for 1–22% of all football injuries (Tysvaer, 1992, Mallo *et al.*, 2011, Boden *et al.*, 1998).

3. Mechanical characteristics and injury risk

As the AT surface developed and improved from 1G to the current 3G, surface characteristics significantly changed and altered the player–surface interaction and consequently the injury risk. Even within the current 3G AT surfaces, the quality depending on the maintenance, age, frequency of use and climate as well as the actual making of the AT depending on the manufacturer varies between sites. Therefore it is of crucial importance to reflect on the surface type being investigated when assessing the impact of artificial surface on injury. Early literature especially (Scranton *et al.*, 1997, DeLee and Farney, 1992, Clarke and Powell, 1979, Ekstrand and Nigg, 1989) reflects on surfaces that are no longer routinely used (Drakos *et al.*, 2013). Therefore, this chapter will focus on evidence evaluating 3G artificial playing surfaces. If an AT fulfills specific quality criteria set by FIFA with respect to mimicking high performance and safety requirements, the AT turf will be marked as a "FIFA quality" (former 1 star) or a "FIFA quality pro" (former 2 star) surface and is then allowed for official match play in the respective (community vs. professional) level. However, unfortunately often the special mechanical properties of the investigated AT surfaces and quality ranking are often not well documented in the research papers and usually the investigated surface is generally referred to as 3G AT. Where available, the quality and surface characteristics were included in this chapter.

Even though the development of the 3G artificial surfaces has addressed problems such as impact absorption, ball bounce and roll, and making up of the infill to more closely replicate the feeling of natural surfaces, an increased risk of non-contact injury when playing on AT is still perceived by the players (Poulos *et al.*, 2014) with the surface being recalled in player's opinions as "too hard", "more abrasive" and "providing less grip" when compared to natural grass (Roberts *et al.*, 2014). Whether this perception is influenced by the historically problematic artificial surfaces or an actual real need for further improvements is currently addressed by various research studies investigating the risk of injury on artificial surfaces and will be discussed in this chapter.

A number of studies have focused on the epidemiology of injury on artificial surfaces suggesting that the overall risk of injury is comparable between artificial turf and natural grass, but that in football the location and type of injury might differ according to the playing surface (Steffen *et al.*, 2007; Ekstrand *et al.*, 2011a; Ekstrand *et al.*, 2006; Fuller *et al.*, 2007b; Fuller *et al.*, 2007a; Soligard *et al.*, 2012; Aoki *et al.*, 2010) affecting either the ankle (Soligard *et al.*, 2012; Ekstrand *et al.*, 2006; Ekstrand *et al.*, 2011b; Ekstrand *et al.*, 2011a), the knee (Fuller *et al.*, 2010), the quadriceps (Steffen *et al.*, 2007), the lower back (Aoki *et al.*, 2010; Soligard *et al.*, 2012), or the spine and shoulder collarbone (Soligard *et al.*, 2012). To receive a better overall view of the effect of AT, these named studies have been the basis of a review of injury incidences by Williams *et al.* (2011) (all studies except Aoki *et al.* (2010), but additionally including Bjorneboe *et al.* (2010)) and a meta-analysis by Williams *et al.* (2013) (all 8 studies). While both author-groups agree on the same likelihood of sustaining an injury on AT as on NT, they resulted in different opinions with respect to sustaining an injury at a specific body location: Williams *et al.* (2011) used the incidence injury rate ratio on AT and NT, with NT as the reference. Clinical inference regarding the true value of effects for each subgroup (e.g. gender, training/match, elite/collegiate, youth/adult) were made according to Batterham and Hopkins (2006) and determined the likelihood that the true magnitude of an effect was substantial in a beneficial or a harmful way. Overall the authors concluded with respect to injury location there was a likely increased risk of ankle injury on AT as they found a likely harmful effect on ankle injuries when playing on AT for five football subgroups (Ekstrand *et al.*, 2011a, Ekstrand *et al.*, 2006, Bjorneboe *et al.*, 2010, Fuller *et al.*, 2007b, Steffen *et al.*, 2007), a trivial effect for two subgroups (Fuller *et al.*, 2007b, Fuller *et al.*, 2007a), and a beneficial effect to reduce ankle injuries for one subgroup (Soligard *et al.*, 2012). With respect to muscle injury, six cohorts showed beneficial effects of AT (Ekstrand *et al.*, 2011a, Steffen *et al.*, 2007, Ekstrand *et al.*, 2006), and no harmful effects were found in the remaining subgroups. Williams *et al.* (2013) took a different approach and computed the unweighted sum of total injuries and exposure times. The resulting adjusted injury rate ratios were statistically analyzed using the Mantel-Haenszel method for fixed effects and categorized based on the conditions of match/training, male/female, and youth/adult. The overall injury rate ratio was calculated over all studies to be 0.86 (injuries on AT/injuries NT). The additional analysis of subgroups (injury location) and conditions (match/training, gender and age) found in many cases an injury rate ratio being below 1 and in no case an injury rate ratio being greater than 1, suggesting that the injury rate on AT is similar or even smaller on AT than on NT. However, the authors draw attention to the considerable heterogeneity between the studies and point out the difficulties of making firm conclusions regarding the influence of AT on player safety. These two analyses highlight the difficulties of drawing an overall conclusion in the grand scheme as the current research situation consists of isolated pockets touching on many aspects possibly influencing the risk of injury (e.g. training/match, gender, level of play, regional aspects, various AT and pitch situations, etc.) and contributing in bits and pieces to an overall picture, that is only in the beginning stages.

Three recent studies further support the use of AT as an alternative safe surface in comparison to NT. Over five seasons Meyers (2013) investigated the injury risk of female football athletes from 13 universities and documented with an injury incidence rate of 7.7, a significantly lower total injury incidence rate compared to NT (9.5 injury incidence rate). Also substantial injuries were significantly lower on AT than on NT (0.7 vs. 1.5 injury incidence rate) and the author concluded that during match games AT and NT are similar with respect to injury rate incidence, but that during training AT is a practical alternative to NT. Almutawa *et al.* (2014) conducted a methodologically interesting but small study investigated the injury rate of Saudi National Team footballers, 32 of which played in the Asian Cup on grass, and 31 in the Gulf Cup on 3G AT and 14 players in both. The authors report a 56.1 per 1000 playing hours (game and training) injury rate for grass and a total of 37.9 injury incidences for 1000 playing hours on AT, suggesting a decreased injury risk on the current 3G turf. While no differences in minor and mild injuries were shown, severe injury had a higher incidence rate on grass (3 vs 1 severe injury per 1000 hours playing exposure). Also lower limb injuries were more likely to be sustained on grass (14.2 vs. 7.9 per 1000 playing hours) than on AT, while muscle injuries showed the same injury rate for both surfaces. Interestingly the risk of sustaining an injury involving player contact was approximately 4 times more likely on grass than on AT (11.5 vs 3.2 injuries per 1000 h playing exposure). O'Kane *et al.* (2015) interviewed 351 American female youth soccer players (11–15 years) regarding injury, field surface, shoe type and position. Injuries occurred with 173 acute injuries mainly at the lower extremity involving primarily the ankle (39.3%), knee (24.9%), and thigh (11.0%). Over half (52.9%) of the injuries were recovered from within one week, whereas 30.2% lasted beyond two weeks. During practices, those injured were approximately 3-fold (OR, 2.83; 95% CI, 1.49-5.31) more likely to play on grass than artificial turf and 2.4-fold (95% CI, 1.03-5.96) more likely to wear cleats on grass than other shoe and surface combinations.

It becomes apparent that reporting injuries due to playing on AT requires a multidimensional view, with various aspects affecting the overall injury risk. This chapter will now aim to describe single aspects of musculoskeletal injuries, such as foot and ankle, knee, lower back and head injury, which playing on AT possibly could influence as well as non-musculoskeletal injuries, such as skin abrasion, infection, carcinogenic risk and airway irritation. Additionally mechanistic aspects and extrinsic factors that may play a role in injury occurrence such as shoe–surface traction, climate, habitational effects and code of sport will be discussed.

4. Musculoskeletal injuries

4.1 Ankle ligament injury

As already stated in section 3, *Mechanical characteristics and injury risk*, the risk of sustaining an ankle injury due to playing on AT is controversially discussed. Ekstrand *et al.* (2006) attribute a 1.81 ratio to the increased rate of ankle sprain injury when performing on AT versus NT, similarly Steffen *et al.* (2007) report a trend towards more ankle sprains on AT in Norwegian female U-17 teams. This increased risk of receiving an ankle sprain injury when playing on AT was accompanied by a decreasing of the likelihood of a quadriceps strain (Ekstrand *et al.*, 2011a; Ekstrand *et al.*, 2006). These findings are in contrast to a study by Soligard *et al.* (2012), who notice a significant lower risk of ankle injuries (ratio 0.5 AT:NT) and (Fuller *et al.*, 2007b; Fuller *et al.*, 2007a), who reported no significant effect on ankle injury when performing on AT.

4.2 Knee injury

In the discussion of injury due to AT, the change in frictional forces and shoe–surface interaction due to the synthetic surface has to be considered as a potentially relevant risk factor to increase the number of ACL-ruptures (Drakos *et al.*, 2010; Torg & Quedenfeld, 1971). Female football players have a 2–3 times higher ACL injury risk compared to their male counterparts (Walden *et al.*, 2011b; Fuller *et al.*, 2007a) and approximately 50% of all season-ending injuries during match play in female football are ACL-tears (Fuller *et al.*, 2007a). In a global perspective Walden *et al.* (2011b) conclude though that ACL injury is a rather unusual football injury rarely constituting more than 5% of all the time loss injuries regardless of playing levels. Balazs *et al.* (2015) reviewed 10 studies with 963 ACL injuries in American football (six studies) and football (soccer) (four studies: Fuller *et al.*, 2007a; Fuller *et al.*, 2007b; Meyers, 2013; Bjorneboe *et al.*, 2010). While in American football four studies found an increased risk of ACL injury on AT, (two included 1G surfaces), only one study identified reduced risk in ACL injury on AT. A different picture seems to occur for football. In the four included studies investigating football players the ACL injury incidence was between 0.31–0.86 (ACL injury incidence rate on turf/ACL injury incidence rate on grass). This ratio indicates that proportionally less ACL injuries occurred on AT than on NT. Supporting these results, a performance analysis of an unanticipated cutting movement of female footballers suggests that the observed kinematic risk factors (knee valgus, knee internal rotation) to sustain an ACL injury are not increased during this change of direction maneuver on AT (Strutzenberger *et al.*, 2014).

4.3 Back, spine and shoulder collarbone injuries

In a cohort of 332 Japanese adolescent soccer players, the incidence of acute contusion/sprain of lumbar spine was with 39–40 injuries per 1000 playing hours similar between the turfs and was the cause of 26% of all injuries occurring in one year on AT and 36% on NT (Aoki *et al.*, 2010). Considering chronic back pain, the group training on AT showed with 42% of all chronic injuries a significantly higher incidence rate for chronic back/low back pain than the group training on NT (33%). As the AT group spent significantly more hours training, early adolescence and prolonged training hours were factors associated with an increased incidence of chronic pain in the AT group (Aoki *et al.*, 2010). This is in line with Soligard *et al.* (2012) who investigated the acute injury risk in a cohort of 60,000 Norwegian adolescent football players over a four-year period. The authors report a higher risk of back and spine (ratio 1.92 AT:NT) and shoulder and collarbone injuries (ratio 2.23) on AT compared with NT.

4.4 Head injuries – concussions

The most common injury mechanisms for a concussion are a collision with another player's head (28%), the ball (24%) or elbow (14%). Ten percent are caused in contact with the ground (Boden *et al.*, 1998). When playing on AT, the risk of sustaining a head injury is approximately doubled (Shorten *et al.*, 2003b; Naunheim *et al.*, 2002). Theobald *et al.* (2010) compared the risk of the head impact causing a mild traumatic brain injury (head-injury-criterion HIC) of six 3G AT with natural grass in dry and wet condition. All six surfaces had a risk of >10% when falling from >77cm independent of wetness, while the grass surface necessitated falls from heights exceeding those achievable during a game to cross the 10% risk margin. The best AT turf performer were surfaces that included a shock pad and a sand and rubber infill, implying that these are key components in impact attenuation. Additionally the HIC seems to be influenced by the surrounding temperature. The investigated 3G AT displayed a HIC of 1.40 ± 0.06 m in temperatures of approximately 15°C, while the HIC significantly decreased to 1.00 ± 0.06 m in temperatures of approximately 2°C (Charalambous *et al.*, 2016). However, if actually more head injuries occur on AT is currently not investigated.

5. Non-musculoskeletal injuries

Besides damage to muscle and bone, AT has been under the spotlight for the concern of human health issues of chemical "contamination" from the materials used in the manufacture and construction of AT and also biological contamination from bacteria that may grow more prevalently in the conducive environment of an artificial pitch (Charalambous *et al.*, 2016). Hence skin abrasion, infection, carcinogenic risk and airway irritation need to be considered.

5.1 Skin abrasion and skin infection

Two factors in respect of artificial surface can contribute to infections. The first is the likelihood of the surface to cause abrasions that can become superinfected and secondly the hospitability of the surface to pathogens (Drakos *et al.*, 2013). Artificial surfaces might facilitate population of the *Staphylococcus aureus* bacteria, which is pathogenic bacterium for strains resistant to methicillin-type antibiotics and can cause *community-associated methicillin-resistant S aureus* (CA-MRSA) infection. In a laboratory setting community-associated MRSA inoculum can survive on AT in high concentrations for one week and in lower concentrations up to a month (Waninger *et al.*, 2011). It is usually harmless and also prevalent on the skin, but can access the internal body through cuts and grazes and cause infection that generally take longer to treat and can be severe if untreated (Fleming, 2011). In a study of American footballers the risk of a CA-MRSA attack rate was 18% in a college team if the player received a turf burn (Begier *et al.*, 2004). While skin abrasion where more common on AT surfaces in the earlier generation surfaces (Ekstrand & Nigg, 1989), in a clinical setting a slide performed on artificial turf caused less erythema but more abrasion compared to natural grass (Peppelman *et al.*, 2013), which is in line with players still being "somewhat dissatisfied" concerning skin abrasions (in sliding tackles) on AT (Burillo *et al.*, 2014). Interestingly in the Soligard *et al.* (2012) report on 60,000 players followed over four years, the occurrence of abrasions and lacerations to be low on both the current 3G AT and NT, showing no significant difference between the surfaces. While another study showed an incident ratio of 0.46 for laceration/skin lesion on AT vs NT in male players in training, but an incident ratio of 1.99 for female players. However both comparisons did not reach significant differences between turfs (Fuller *et al.*, 2007b). As skin abrasions are minor injuries not influencing the game, they are often not covered by epidemiological data (Peng Tay *et al.*, 2015) and hence more research in this area is needed.

5.2 Carcinogenic risk due to rubber infill

The rubber infill (recycled from tires) has in some countries been restricted in its use, though the increasing volume of literature suggests a very low risk to human health from contact or inhalation (Fleming, 2011). While Drakos *et al.* (2013) states that concerns that the rubber infill could, through inhalation, ingestion or topical contact, set a carcinogenic risk, have not been substantiated, the EHHI warns that toxic and carcinogenic chemicals were found in rubber infills used for AT and playgrounds (http://www.ehhi.org/turf/turf_pr_0810.shtml). This highlights that the last word on carcinogenic risk has not yet been spoken. The Environment and Human Health Inc., (EHHI) refers to a cancer list collected by Amy Griffin, Associate Head Soccer Coach at the University of Washington, which counts cancer cases among athletes, who have played on synthetic turf fields. Of the 234 reported cancer cases, 186 are soccer players with 116 being soccer goalkeepers (http://www.ehhi.org/turf-cancer-stats.php). "One would expect the soccer goalies to be the first to be affected because they are the ones who are always diving into the fields and therefore are the most exposed to the carcinogens in the crumb rubber" explains the EHHI (http://www.prnewswire.com/news-releases/the-cancer-list-keeps-growing-among-athletes-on-synthetic-turf-300099336.html).

5.3 Airway irritation

No published data suggest that modern third-generation synthetic playing surface is a source of allergens that may potentiate allergic rhinitis or exacerbate asthma (Drakos *et al.*, 2013).

6. Extrinsic factors influencing injury risk

6.1 Traction: Shoe–surface interface

Outdoor football shoes are equipped with blades/cleats or studs to penetrate and interlock with the playing surface depending on surface hardness. The sport shoe industry provides the athlete with a variety of possible shoe designs to best meet the requirements of the specific surface such as natural soft grip, firm grip, and AT. Cleats are made of elastomeric material or thermoplastic polyurethane or steel-tipped thermoplastic polyurethane (Villwock *et al.*, 2009) and come in varying shapes (edge-type, bladed, conical, cup-shaped, tapered, triangular, or elliptical (Culpepper & Niemann, 1983). To play on AT a *turf shoe* is often used, which has a dense pattern of short (6.5 mm) elastomeric studs distributed over the entire sole (Figure 17.1.a, usually used by recreational level players) (e.g. Drakos *et al.*, 2013; Villwock *et al.*, 2009). This is in contrast to conventional football shoes that serve according to the surface characteristics. For example, the shoe displayed in Figure 17.1.b consisting of more than 12 studs in various lengths and cleat tip diameters is used for firm and dry grounds, but also might be played on AT, while the shoe displayed in Figure 17.1.c. is a *hybrid shoe* (also called mixed outsole shoe) combining cleats with long and short studs and is often chosen for soft (and wet) grounds; parallel to the hybrid shoe, purely soft ground shoes, having usually six iron studs, are mainly used by goalkeepers and defenders.

During movement the interlock between playing surface and shoe generates traction and influences the player's ability to accelerate, decelerate or to change direction (Fleming, 2011). Severn (2010) identified four primary factors affecting the complex behavior of shoe–surface traction: the sports specific movement (mass and speed of athlete, angle of foot, loading rate and height before contact), the footwear (the characteristics of the studs such as number, size, shape and material, as well as sole material and contact surface area), the playing surface (physical and mechanical characteristics of the carpet and infill layer and shockpad thickness), and the environment (such as water, temperature, chemicals and maintenance).

The level of traction produced at the shoe–surface interface is important as a too-low traction resistance force will cause slipping, an optimal zone of traction will allow for optimal performance and may minimize injury risk (Luo & Stefanyshyn, 2011, Smeets *et al.*, 2012, Nigg & Segesser, 1988), while excessive traction resistance will cause foot sticking (Shorten *et al.*, 2003a) and high levels of rotational traction are identified as contributor to injury (Thomson *et al.*, 2015). It has been suggested that possible measures to reduce shoe–surface traction, such as ground watering and softening, ply during the winter months, use of natural grasses, use of boots with shorter cleats, could all reduce the risk of injury (Orchard, 2002). To identify a stringent injury mechanism seems quite difficult, though many authors have investigated traction behavior mechanically, using a variety of shoe and surface types, concluding that the traction generated at the shoe–surface interface is dependent on each shoe–surface combination (Cawley *et al.*, 2003, Villwock *et al.*, 2009, Severn *et al.*, 2011). However, the level or representativeness of these mechanical tests for the human–surface interaction

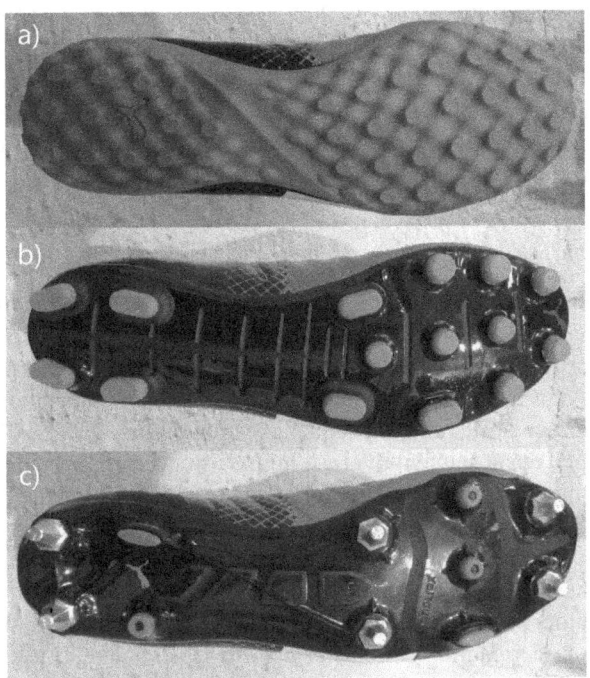

Figure 17.1 Examples of various types of football shoes a) artificial turf shoe, b) firm ground shoe, c) soft ground shoe

has been questioned (Fleming, 2011; Fleming *et al.*, 2005). For example, while measures by the Artificial Athlete imply a high correlation between impact force and surface stiffness, measures of the human–surface ground reaction force have revealed no correlation between the peak impact force and the stiffness of the surface (Nigg & Yeadon, 1987; Nigg, 1990; Dixon *et al.*, 1999). Another example in a comparative laboratory setting showed no differences in peak torque and rotational stiffness between bladed cleats, molded rubber studs and metal studs on AT and NT (Galbusera *et al.*, 2013), but injured American female youth soccer players were during training 2.4-fold more likely to having played on grass with cleats when they were injured compared to having played with cleats on AT or any other shoe and surface (AT, NT) combination (O'Kane *et al.*, 2015).

Two reviews (Taylor *et al.*, 2012; Drakos *et al.*, 2013) identified shoe type (Heidt *et al.*, 1996; Cawley *et al.*, 2003; Villwock *et al.*, 2009), sole material (Villwock *et al.*, 2009), pattern of cleats (Andreasson *et al.*, 1986; Lambson *et al.*, 1996), and shape of cleats (Lambson *et al.*, 1996; Queen *et al.*, 2008; Bonstingl *et al.*, 1975) as possible influences on the amount of torque or rotation stiffness being developed at the shoe–surface interface. They concluded higher torques exist when cleated shoes are used on AT being related to the greater total effective area involved in cleat–surface contact. The total effective contact area is proportional to cleat number, length and size

(Cawley *et al.*, 2003; Torg *et al.*, 1974; Torg & Quedenfeld, 1971; Bonstingl *et al.*, 1975). This is in agreement with Villwock *et al.* (2009), reporting that AT yielded significantly higher peak torque and rotational stiffness than the NT on all investigated shoe designs. The turf shoe showed on all surfaces (AstroPlay, 3G AT, NT-sand based, NT-native soil) the lowest peak torque. Authors also highlight that due to the large number of cleat patterns, materials and sizes and constant changes in the athletic shoe market, long-term data and well-designed prospective studies are scarce.

6.2 Injury due to surface shifts

Frequent shifts between surfaces have been speculated as possible risk factor (Ekstrand & Nigg, 1989; Kristenson *et al.*, 2013). The recent study by Kristenson *et al.* (2015) now firstly provided evidence data on this issue for 32 male top-division clubs in Scandinavian footballs: The authors did not find an association between the number of surface shifts and subsequent overuse injury risk (1.01 risk ratio). Also no difference in subsequent overuse injury risk occurred whether the match was on accustomed similar surface to home venue (AT-AT or NT-NT) or unaccustomed surface (1.04 risk ratio). If the home venue had a grass surface installed, athletes were less likely to sustain a match injury when the away match was played on an AT surface (0.66 injury rate). Whether this is originated in a psychological twist, that actors believe in a higher injury risk when playing on AT and therefore played with higher caution (Poulos *et al.*, 2014), a change in team tactics or an actual benefit of the surface cannot be answered. Overall the authors suggest though, that clubs and players can cope with such surface transitions.

6.3 Injury due to playing hours

Aoki *et al.* (2010) identified a higher risk for chronic lower back pain in adolescent players when mainly training on AT. The authors speculate that the prolonged training time registered for the group training on AT could have potentially provoked this effect and suggest close monitoring. Underpinning this result Kristenson *et al.* (2013) report that while surface type did not affect the injury incidence at individual player level, clubs with AT as their home venue surface showed higher incidence rates of acute training injuries and overuse injuries compared with clubs playing home matches on NT.

6.4 Climate

The climate is under scrutiny due to two effects: 1) the temperature can affect the ground hardness and shoe–surface traction on NT, with the shoe–surface hardness increasing on harder and drier grounds (Orchard, 2002) and hence possibly influence the injury risk; and 2) it has been reported that AT heats up substantially more in hot temperatures, possibly due to a better absorption of solar radiation (Williams & Pulley, 2003, Thoms *et al.*, 2014, McNittt *et al.*, 2008, Petrass *et al.*, 2014). This could lead to, for example, blisters and cuts (Bossi, 2015, Williams &

Pulley, 2003), physiological responses (Simon, 2010), as well as an increased of-gassing of possible harmful chemical particles (EHHI, 2010). With respect to the first aspect: in laboratory testing, the impact attenuation properties of AT were found to be independent of moisture content (dry and wet), while NT varied depending on its usage (Theobald *et al.*, 2010). With respect to the second aspect, when the surface temperature reaches above 44°C, cutaneous thermal injury can occur, temperatures above 77°C can lead to second degree burns already after 35 seconds of exposure (Harrington *et al.*, 1995). However, no epidemiological study exists reporting these injuries (blisters, heat burns etc.), possibly due to their minor influence of the quality of the game or training, not covered in the two most common injury descriptions, and might hence share the same faith as the report of skin abrasion. Across varying weather conditions, the more consistent nature of AT could however be beneficial, but limited research is available for actual football play (Williams *et al.*, 2011). It may well be that other aspects affecting injury risk than the temperature might be of more importance, as an example from American football shows. In 2004 Meyers and Barnhill (2004) conclude a higher injury risk for games being played during temperatures >70°F (or 21.1°C), while a later study quite contradictorily reported that injury incidence was substantially lower on hot days (>70°F) (Meyers, 2010), with a different artificial surface used. The authors concluded that the condition of the pitch during these temperatures might have a substantial influence. Current research develops predictive models on the surface heat, which could provide better planning information (Peng Tay *et al.*, 2015), and investigates which parts of the AT absorb the heat the most, in order to further improve AT technology (Petrass *et al.*, 2014).

6.5 Code of sport

Rugby, football (soccer) and American football are all played on AT, but as playing styles differ, different injury incidence patterns on different surfaces might occur. As such for example, research investigating the surface effect on injury in the National Football League reports a 22% higher risk for each knee and ankle sprain on AT respectively. In detail the risk for receiving an ACL strain was 67% higher and the risk for receiving an ankle eversion sprain was 31% higher on AT (Hershman *et al.*, 2012). In Australian rules football (AFL) differences for ACL injury risk were found for different species of grass (Orchard *et al.*, 2005). Fuller *et al.* (2010) notice that, even though statistically not significant, rugby players were four times more likely to receive an ACL injury on AT (ratio 3.82 AT:NT) contradicting the reported results for soccer. As AT is advertised as a multipurpose surface, the different needs of the varying sports performed on AT would need to be further investigated and considered in turf construction.

7. Conclusion

This chapter has explored three main areas, namely musculoskeletal injury, non-musculoskeletal injury and extrinsic factors that may influence injury risk on artificial turf during football. From the current body of research it was identified that AT does not create an increase in musculoskeletal injury at the specific sites of the foot, ankle and knee. However, there is some evidence to suggest that due to the changes in the mechanical characteristics of the surface stiffness, chronic back pain and head injury would be an issue, although epidemiological evidence does not sufficiently exist yet. This chapter has identified that the non-musculoskeletal components including skin abrasion and infection, carcinogenic risk and airway irritation have little influence over injury risk, however there is a shortage of evidence that can be drawn upon in these key areas and this is a future direction for turf research.

The final section made specific observations relating to the extrinsic factors that may influence injury risk on AT. Of these, it was identified that the shoe–surface interface does not increase injury risk. It is speculated that the advantage of prolonged training on artificial surfaces might lead to an increased risk of chronic injury, however, there was little evidence to suggest that changing between artificial and natural turf was an etiological factor. Issues of climate and football code in the chapter highlighted a gap in the scientific literature, which must be attended to. The current body of research implies no increased risk of injury when playing on AT.

This review demonstrates the need to increase our understanding of the mechanisms that influence injury risk; the epistemological approach chosen must be one that demonstrates at the very least a multi- and ultimately an inter-disciplinary perspective. The challenge for scientists is to combine the need for uniformity of playing surfaces with injury, economic and social requirements.

References

Agel, J., Evans, T. A., Dick, R., Putukian, M. & Marshall, S. W. (2007). Descriptive epidemiology of collegiate men's soccer injuries: National Collegiate Athletic Association Injury Surveillance System, 1988–1989 through 2002–2003. *Journal of Athletic Training*, 42, 270–277.

Almutawa, M., Scott, M., George, K. P. & Drust, B. (2014). The incidence and nature of injuries sustained on grass and 3rd generation artificial turf: a pilot study in elite Saudi National Team footballers. *Physical Therapy in Sport*, 15, 47–52.

Andreasson, G., Lindenberger, U., Renstrom, P. & Peterson, L. (1986). Torque developed at simulated sliding between sport shoes and an artificial turf. *The American Journal of Sports Medicine*, 14, 225–230.

Aoki, H., Kohno, T., Fujiya, H., Kato, H., Yatabe, K., Morikawa, T. & Seki, J. (2010). Incidence of injury among adolescent soccer players: a comparative study of artificial and natural grass turfs. *Clinical Journal of Sport Medicine*, 20, 1–7.

Balazs, G.C., Pavey, G. J., Brelin, A. M., Pickett, A., Keblish, D. J. & Rue, J. P. (2015). Risk of anterior cruciate ligament injury in athletes on synthetic playing surfaces: A systematic review. *The American Journal of Sports Medicine*, 43, 1798–804.

Batterham, A. M. & Hopkins, W. G. (2006). Making meaningful inferences about magnitudes. *International Journal of Sports Physiology and Performance*, 1, 50–57.

Begier, E. M., Frenette, K., Barrett, N. L., Mshar, P., Petit, S., Boxrud, D. J., Watkins-Colwell, K., Wheeler, S., Cebelinski, E. A., Glennen, A., Nguyen, D. & Hadler, J. L. (2004). A high-morbidity outbreak of methicillin-resistant Staphylococcus aureus among players on a college football team, facilitated by cosmetic body shaving and turf burns. *Clinical Infectious Diseases*, 39, 1446–1453.

Bengtsson, H., Ekstrand, J. & Hagglund, M. (2013). Muscle injury rates in professional football increase with fixture congestion: An 11-year follow-up of the UEFA Champions League injury study. *British Journal of Sports Medicine*, 47, 743–747.

Bjorneboe, J., Bahr, R. & Andersen, T. E. (2010). Risk of injury on third-generation artificial turf in Norwegian professional football. *British Journal of Sports Medicine*, 44, 794–798.

Boden, B. P., Kirkendall, D. T. & Garrett, W. E. Jr. (1998). Concussion incidence in elite college soccer players. *The American Journal of Sports Medicine*, 26, 238–241.

Bonstingl, R. W., Morehouse, C. A. & Niebel, B. W. (1975). Torques developed by different types of shoes on various playing surfaces. *Medicine and Science in Sports*, 7, 127–131.

Bossi, D. (2015). Matildas say synthetic turf 'like hot coals. *Sydney Morning Herald*.

Burillo, P., Gallardo, L., Felipe, J. L. & Gallardo, A. M. (2014). Artificial turf surfaces: perception of safety, sporting feature, satisfaction and preference of football users. *European Journal of Sports Science*, 14 Suppl 1, S437–447.

Cawley, P. W., Heidt, R. S. Jr., Scranton, P. E. Jr., Losse, G. M. & Howard, M. E. (2003). Physiologic axial load, frictional resistance, and the football shoe–surface interface. *Foot & Ankle International*, 24, 551–556.

Charalambous, L., von Lieres und Wilkau, H., Potthast, W. & Irwin, G. (2016). The effects of artificial surface temperature on mechanical properties and player kinematics during landing and acceleration. *Journal of Sport and Health Science*, 5, 355–360.

Clarke, K. S. & Powell, J. W. (1979). Football helmets and neurotrauma–an epidemiological overview of three seasons. *Medicine and Science in Sports*, 11, 138–145.

Culpepper, M. I. & Niemann, K. M. (1983). An investigation of the shoe–turf interface using different types of shoes on Poly-Turf and Astro-Turf: torque and release coefficients. *The Alabama Journal of Medical Sciences*, 20, 387–390.

Delee, J. C. & Farney, W. C. (1992). Incidence of injury in Texas high school football. *The American Journal of Sports Medicine*, 20, 575–580.

Dick, R., Putukian, M., Agel, J., Evans, T. A. & Marshall, S. W. (2007). Descriptive epidemiology of collegiate women's soccer injuries: National Collegiate Athletic Association Injury Surveillance System, 1988–1989 through 2002–2003. *Journal of Athletic Training*, 42, 278–285.

Drakos, M. C., Hillstrom, H., Voos, J. E., Miller, A. N., Kraszewski, A. P., Wickiewicz, T. L., Warren, R. F., Allen, A. A. & O'Brien, S. J. (2010). The effect of the shoe–surface interface in the development of anterior cruciate ligament strain. *Journal of Biomechanical Engineering*, 132(1), 011003, doi: 10.1115/1.4000118.

Drakos, M. C., Taylor, S. A., Fabricant, P. D. & Haleem, A. M. (2013). Synthetic playing surfaces and athlete health. *Journal of the American Academy of Orthopaedic Surgeons*, 21, 293–302.

Dvorak, J. & Junge, A. (2000). Football injuries and physical symptoms. A review of the literature. *The American Journal of Sports Medicine*, 28, S3-9.

Ekstrand, J., Hagglund, M. & Fuller, C. W. (2011a). Comparison of injuries sustained on artificial turf and grass by male and female elite football players. *Scandinavian Journal of Medicine & Science in Sports*, 21, 824–832.

Ekstrand, J., Hagglund, M. & Walden, M. (2011b). Epidemiology of muscle injuries in professional football (soccer). *The American Journal of Sports Medicine*, 39, 1226–1232.

Ekstrand, J. & Nigg, B. M. (1989). Surface-related injuries in soccer. *Sports Medicine*, 8, 56–62.

Ekstrand, J., Timpka, T. & Hagglund, M. (2006). Risk of injury in elite football played on artificial turf versus natural grass: a prospective two-cohort study. *British Journal of Sports Medicine*, 40, 975–980.

Elliot, B., Hershman, E. B., Anderson, R., Bergfeld, J. A., Bradley, J. P., Coughlin, M. J., Johnson, R. J., Spindler, K. P., Wojtys, E., Powell, J. W. & the National Football League Injury and Safety Panel (2012). An analysis of specific lower extremity injury rates on grass and Field Turf playing surfaces in National Football League Games: 2000–2009 seasons. *The American Journal of Sports Medicine*, 40, 2200–2205.

Environmental and Human Health, Inc. (EHHI). Artificial Turf: Cancers Among Players. Internet page, http://www.ehhi.org/turf-cancer-stats.php, retrieved 12.7.2017)

Environmental and Human Health, Inc. (EHHI). (2015). The Cancer List Keeps Growing Among Athletes on Synthetic Turf. Internet page, http://www.prnewswire.com/news-releases/the-cancer-list-keeps-growing-among-athletes-on-synthetic-turf-300099336.html, retrieved 12.7.2017

Eurosafe (2013). Injuries in the European Union, Report on injury statistics 2008–2010. Amsterdam: European Association for Injury Prevention and Safety Promotion (Eurosafe).

Fédération Internationale de Football Association (2007). FIFA Big Count 2006: 270 million people active in football. In: *SERVICES, F. C. D. I.* Zurich, Switzerland.

Fleming, P. (2011). Artificial turf systems for sport surfaces: Current knowledge and research needs. *Journal of Sports Engineering and Technology*, 225, 43–63.

Fleming, P. R., Young, C., Roberts, J. R., Jones, R. L. & Dixon, N. (2005). Human perceptions of artificial surfaces for field hockey. *Sports Engineering*, 8, 121–136.

Fuller, C. W., Clarke, L. & Molloy, M. G. (2010). Risk of injury associated with rugby union played on artificial turf. *Journal of Sports Sciences*, 28, 563–570.

Fuller, C. W., Dick, R. W., Corlette, J. & Schmalz, R. (2007a). Comparison of the incidence, nature and cause of injuries sustained on grass and new generation artificial turf by male and female football players. Part 1: match injuries. *British Journal of Sports Medicine*, 41 Suppl 1, i20–26.

Fuller, C. W., Dick, R. W., Corlette, J. & Schmalz, R. (2007b). Comparison of the incidence, nature and cause of injuries sustained on grass and new generation artificial turf by male and female football players. Part 2: training injuries. *British Journal of Sports Medicine*, 41 Suppl 1, i27–32.

Fuller, C. W., Ekstrand, J., Junge, A., Andersen, T. E., Bahr, R., Dvorak, J., Hagglund, M., Mccrory, P. & Meeuwisse, W. H. (2006). Consensus statement on injury definitions and data collection procedures in studies of football (soccer) injuries. *British Journal of Sports Medicine*, 40, 193–201.

Galbusera, F., Tornese, D. Z., Anasetti, F., Bersini, S., Volpi, P., La Barbera, L. & Villa, T. (2013). Does soccer cleat design influence the rotational interaction with the playing surface? *Sports Biomechanics*, 12, 293–301.

Hagglund, M., Walden, M., Bahr, R. & Ekstrand, J. (2005). Methods for epidemiological study of injuries to professional football players: developing the UEFA model. *British Journal of Sports Medicine*, 39, 340–346.

Harrington, W. Z., Strohschein, B. L., Reedy, D., Harrington, J. E. & Schiller, W. R. (1995). Pavement temperature and burns: streets of fire. *Annuals of Emergency Medicine*, 26, 563–568.

Heidt, R. S. Jr., Dormer, S. G., Cawley, P. W., Scranton, P. E. Jr., Losse, G. & Howard, M. (1996). Differences in friction and torsional resistance in athletic shoe–turf surface interfaces. *The American Journal of Sports Medicine*, 24, 834–842.

Hershman, E. B., Anderson, R., Bergfeld, J. A., Bradley, J. P., Coughlin, M. J., Johnson, R. J., Spindler, K. P., Wojtys, E., Powell, J. W. & the National Football League Injury and Safety Panel (2012). An analysis of specific lower extremity injury rates on grass and Field Turf playing surfaces in National Football League Games: 2000–2009 seasons. *The American Journal of Sports Medicine*, 40, 2200–2205.

Kristenson, K., Bjorneboe, J., Walden, M., Andersen, T. E., Ekstrand, J. & Hagglund, M. (2013). The Nordic Football Injury Audit: Higher injury rates for professional football clubs with third-generation artificial turf at their home venue. *British Journal of Sports Medicine*, 47, 775–781.

Kristenson, K., Bjorneboe, J., Walden, M., Ekstrand, J., Andersen, T. E. & Hagglund, M. (2016). No association between surface shifts and time-loss overuse injury risk in male professional football. *Journal of Science and Medicine in Sport*, 19, 218–221.

Lambson, R. B., Barnhill, B. S. & Higgins, R. W. (1996). Football cleat design and its effect on anterior cruciate ligament injuries. A three-year prospective study. *The American Journal of Sports Medicine*, 24, 155–159.

Luo, G. & Stefanyshyn, D. (2011). Identification of critical traction values for maximum athletic performance. *Footwear Science*, 33, 127–138.

Majewski, M., Habelt, S. & Steinbrück, K. (2006). Epidemiology of athletic knee injuries. *The Knee*, 13, 184–188.

Mallo, J., Gonzalez, P., Veiga, S. & Navarro, E. (2011). Injury incidence in a Spanish sub-elite professional football team: A prospective study during four consecutive seasons. *Journal of Sports Science and Medicine*, 10, 731–736.

McNittt, A., Petrunak, D. M. & Serensits, T. J. (2007). Temperature amelioration of synthetic turf surfaces through irrigation. In J. C. Stier (ed.) *Proceedings 2nd International Conference on Turfgrass Science and Management for Sport Fields*. (pp. 573–581), Leuven: International Society for Horticultural Science..

Meyers, M. C. (2010). Incidence, mechanisms, and severity of game-related college football injuries on FieldTurf versus natural grass: a 3-year prospective study. *The American Journal of Sports Medicine*, 38, 687–697.

Meyers, M. C. (2013). Incidence, mechanisms, and severity of match-related collegiate women's soccer injuries on FieldTurf and natural grass surfaces: A 5-year prospective study. *The American Journal of Sports Medicine*, 41, 2409–2420.

Meyers, M. C. & Barnhill, B. S. (2004). Incidence, causes, and severity of high school football injuries on FieldTurf versus natural grass: A 5-year prospective study. *The American Journal of Sports Medicine*, 32, 1626–1638.

Naunheim, R., Mcgurren, M., Standeven, J., Fucetola, R., Lauryssen, C. & Deibert, E. (2002). Does the use of artificial turf contribute to head injuries? *Journal of Trauma-Injury Infection & Critical Care*, 53, 691–694.

Nielsen, A. B. & Yde, J. (1989). Epidemiology and traumatology of injuries in soccer. *The American Journal of Sports Medicine*, 17, 803–807.

Nigg, B. M. & Segesser, B. 1988. The influence of playing surfaces on the load on the locomotor system and on football and tennis injuries. *Sports Medicine*, 5, 375–385.

O'Kane, J. W., Gray, K. E., Levy, M. R., Neradilek, M., Tencer, A. F., Polissar, N. L. & Schiff, M. A. (2016). Shoe and field surface risk factors for acute lower extremity injuries among female youth soccer players. *Clinical Journal of Sport Medicine*. doi: 10.1097/JSM.0000000000000236.

Orchard, J. (2002). Is there a relationship between ground and climatic conditions and injuries in football? *Sports Medicine*, 32, 419–432.

Orchard, J. W., Chivers, I., Aldous, D., Bennell, K. & Seward, H. (2005). Rye grass is associated with fewer non-contact anterior cruciate ligament injuries than Bermuda grass. *British Journal of Sports Medicine*, 39, 704–709.

Tay, S. P., Fleming, P. & Forrester, S. (2015). Insights to skin–turf friction as investigated using the Securisport. *Procedia Engineering*, 112, 320–235.

Peppelman, M., van den Eijnde, W. A., Langewouters, A. M., Weghuis, M. O. & van Erp, P. E. (2013). The potential of the skin as a readout system to test artificial turf systems: Clinical and immunohistological effects of a sliding on natural grass and artificial turf. *International Journal of Sports Medicine*, 34, 783–788.

Petrass, L. A., Twomey, A. & Harvey, J. T. (2014). Understanding how the components of a synthetic turf system contribute to increased surface temperature. *Procedia Engineering*, 72, 943–948.

Poulos, C. C., Gallucci, J., Jr., Gage, W. H., Baker, J., Buitrago, S. & Macpherson, A. K. (2014). The perceptions of professional soccer players on the risk of injury from competition and training on natural grass and 3rd generation artificial turf. *BMC Sports Science, Medicine and Rehabilitation*, 6, 11. 1–7.

Queen, R. M., Charnock, B. L., Garrett, W. E., Jr., Hardaker, W. M., Sims, E. L. & Moorman, C. T. 3rd. (2008). A comparison of cleat types during two football-specific tasks on FieldTurf. *British Journal of Sports Medicine*, 42, 278–284.

Roberts, J., Osei-Owusu, P., Harland, A., Owen, A. & Smith, A. (2014). Elite football players' perceptions of football turf and natural grass surface properties. *Procedia Engineering*, 72, 907–912.

Scranton, P. E. Jr., Whitesel, J. P., Powell, J. W., Dormer, S. G., Heidt, R. S. Jr., Losse, G. & Cawley, P. W. (1997). A review of selected noncontact anterior cruciate ligament injuries in the National Football League. *Foot Ankle International*, 18, 772–776.

Severn, K. A. (2010). Player surface interactions: traction on artificial turf. PhD Thesis, Loughborough University.

Severn, K. A., Fleming, P. R., Clarke, J. D. & Carré, M. J. (2011). Science on synthetic turf surfaces: Investigating traction behaviour. In Proceedings of the Institution of Mechanical Engineers, Part P: *Journal of Sports Engineering and Technology*, 225, 147–158.

Shorten, M., Hudson, B. & Himmeslbach, J. (2003a). Shoe–surface traction of conventional and in-filled synthetic turf football surfaces. In: (Eds.), *Proceedings of 15th International Congress of Biomechanics* (p. 6), 11 July 2013 Dunedin, New Zealand. University of Otago.

Shorten, M. R., Bahr, R. & Andersson, T. E. (2003b). Sport surfaces and the risk of traumatic brain injury. In B. M. Nigg, C. K. Cole, & D. Stefanyshyn (eds.) *Sport surfaces*. Calgary, Canada: University of Calgary.

Silvers-Granelli, H., Mandelbaum, B., Adeniji, O., Insler, S., Bizzini, M., Pohlig, R., Junge, A., Snyder-Mackler, L. & Dvorak, J. (2015). Efficacy of the FIFA 11+ Injury Prevention Program in the Collegiate Male Soccer Player. *The American Journal of Sports Medicine*, 43, 2628–2637.

Simon, R. (2010). Review of the impacts of crumb and rubber in artificial turf applications. *Green Manufacutring and Sustainable Manufacturing Partnership*. Berkeley, CA: University of California, Laboratory for Manufacturing and Sustainability.

Smeets, K., Jacobs, P., Hertogs, R., Luyckx, J. P., Innocenti, B., Corten, K., Ekstrand, J. & Bellemans, J. (2012). Torsional injuries of the lower limb: An analysis of the frictional torque between different types of football turf and the shoe outsole. *British Journal of Sports Medicine*, 46, 1078–1083.

Soligard, T., Bahr, R. & Andersen, T. E. (2012). Injury risk on artificial turf and grass in youth tournament football. *Scandinavian Journal of Medicine & Science in Sports*, 22, 356–361.

Steffen, K., Andersen, T. E. & Bahr, R. (2007). Risk of injury on artificial turf and natural grass in young female football players. *British Journal of Sports Medicine*, 41 Suppl 1, i33–37.

Strutzenberger, G., Cao, H.-M., Koussev, J., Potthast, W. & Irwin, G. (2014). Effect of turf on the cutting movement of female football players, *Journal of Sport and Health Science*, 3, 314–319.

Taylor, S. A., Fabricant, P. D., Khair, M. M., Haleem, A. M. & Drakos, M. C. (2012). A review of synthetic playing surfaces, the shoe–surface interface, and lower extremity injuries in athletes. *The Physician and Sportsmedicine*, 40, 66–72.

Theobald, P., Whitelegg, L., Nokes, L. D. & Jones, M. D. (2010). The predicted risk of head injury from fall-related impacts on to third-generation artificial turf and grass soccer surfaces: a comparative biomechanical analysis. *Sports Biomechanics*, 9, 29–37.

Thoms, A. W., Brosnan, J. T., Zidek, J. M. & Sorochan, J. C. (2014). Models for prediciting surface temperatures on synthetic turf playing surfaces. *Procedia Engineering*, 72, 895–900.

Thomson, A., Whiteley, R. & Bleakley, C. (2015). Higher shoe–surface interaction is associated with doubling of lower extremity injury risk in football codes: A systematic review and meta-analysis. *British Journal of Sports Medicine*, 49, 1245–1252.

Torg, J. S. & Quedenfeld, T. (1971). Effect of shoe type and cleat length on incidence and severity of knee injuries among high school football players. *Research Quarterly. American Association for Health, Physical Education and Recreation*, 42, 203–211.

Torg, J. S., Quedenfeld, T. C. & Landau, S. (1974). The shoe–surface interface and its relationship to football knee injuries. *Journal of Sports Medicine*, 2, 261–269.

Tysvaer, A. T. (1992). Head and neck injuries in soccer. Impact of minor trauma. *Sports Medicine*, 14, 200–213.

Villwock, M. R., Meyer, E. G., Powell, J. W., Fouty, A. J. & Haut, R. C. (2009). Football playing surface and shoe design affect rotational traction. *The American Journal of Sports Medicine*, 37, 518–525.

Walden, M., Hagglund, M. & Ekstrand, J. (2005). Injuries in Swedish elite football – a prospective study on injury definitions, risk for injury and injury pattern during 2001. *Scandinavian Journal of Medicine & Science in Sports*, 15, 118–125.

Walden, M., Hagglund, M., Magnusson, H. & Ekstrand, J. (2011a). Anterior cruciate ligament injury in elite football: a prospective three-cohort study. *Knee Surgery, Sports Traumatology, Arthroscopy*, 19, 11–19.

Walden, M., Hagglund, M., Werner, J. & Ekstrand, J. (2011b). The epidemiology of anterior cruciate ligament injury in football (soccer): a review of the literature from a gender-related perspective. *Knee Surgery, Sports Traumatology, Arthroscopy*, 19, 3–10.

Waninger, K. N., Rooney, T. P., Miller, J. E., Berberian, J., Fujimoto, A. & Buttaro, B. A. (2011). Community-associated methicillin-resistant Staphylococcus aureus survival on artificial turf substrates. *Medicine and Science in Sports and Exercise*, 43, 779–784.

Williams, C. F. & Pulley, G. E. (2003). *Synthetic surface heat studies*. Brigham: Brigham Young University.

Williams, J. H., Akogyrem, E. & Williams, J. R. (2013). A Meta-Analysis of Soccer Injuries on Artificial Turf and Natural Grass. *Journal of Sports Medicine* Article ID 380523.

Williams, S., Hume, P. A. & Kara, S. (2011). A review of football injuries on third and fourth generation artificial turfs compared with natural turf. *Sports Medicine*, 41, 903–923.

18 Artificial turf in football

A performance perspective

Gerda Strutzenberger, Wolfgang Potthast and Gareth Irwin

1. Introduction

With the drive by FIFA to globalize the use of artificial turf (AT) in football, the issues surrounding the impact on injury and performance become pertinent areas to explore. This chapter will examine the current research base that has investigated issues associated with this playing surface in terms of the changes to player performance. Starting with a brief history of the introduction of artificial turf, the chapter will then allocate the existing literature to four main foci: firstly, mechanical surface characteristics and performance, secondly biomechanical aspects of player performance, thirdly, the environmental factor of temperature and finally, physiological responses. The terms synthetic turf/surface, artificial turf/surface, football turf/surface are interchangeably used for artificial playing surfaces and will be referred to as artificial turf (AT) within this context.

2. History

The first artificial soccer turf systems consisted of carpet-like systems with synthetic fibers of about 1 cm length. In the 1970s, those kinds of systems were installed simply on concrete or similar surfaces without cushioning structures or substrate between the fibers to allow soccer cleats to penetrate. The next systems had longer fibers and were filled with sand. Both generations were not well accepted by soccer players and were not a successful alternative to natural grass turfs. At about the turn of the millennium, so called third generation (3G) artificial soccer turf systems were launched. They include cushioning structures between the AT fleece and the rigid basis and also contain substrate between the AT fibers allowing for penetration of the cleats and provide adequate traction. The fiber length, therefore, increased in those systems up to about seven centimeters. A usual 3G AT field is installed on a very rigid substrate (e.g. concrete), which is covered by an elastic layer or shock pad of varying materials as foams, rubber composites etc. (Figure 18.1). The purpose of this deformable layer is to reduce impact-related forces and energy restitution. The actual AT fibers are usually tufted in a fleece serving as liner for the fibers. They are classically straight extruded or crimped monofilaments of polyethylene. Some fiber types contain two material components as stiffer cores to keep the AT fiber straight and a

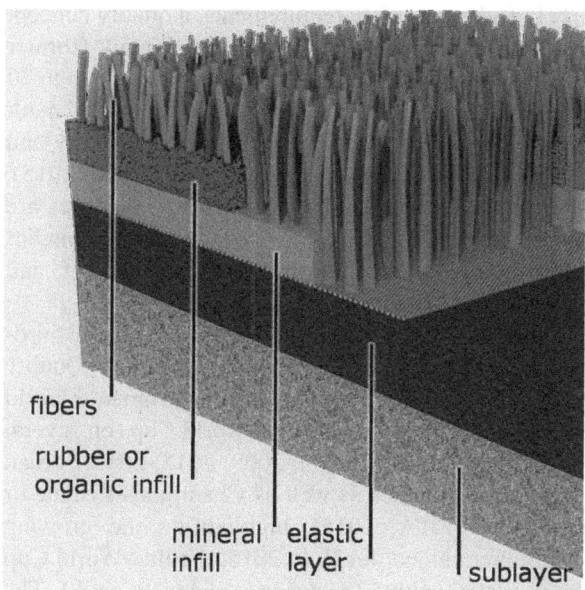

Figure 18.1 A typical construction of third generation artificial turf

different coating material. The tufting of the fibers usually creates a given pile direction of the fibers and therefore might cause a predominant direction of ball rolling and special perception at foot–ground contact. Weaving as a recent development tries to avoid predominant fiber directions and related effects. The space between the fibers is filled with infill material (Figure 18.1). The first purpose of the infill is to weigh the AT carpet down in order to keep it flat on the ground and to support the fibers and preventing them from excessive bending. Usually mineral infill (e.g. sand) is used for this purpose. Furthermore, an elastic infill component is added to allow for penetrating of soccer cleats and providing additional shock absorption. This infill component is usually granulated synthetic material as styrene-butadiene rubber (SBR), polyurethane (PU), Ethylen-Propylen-Dien-Caoutchouc (EPDM) and thermoplastic elastomers. The diameter of the granulate particles (0.5–3 mm) as well as the material itself determines the behavior of the infill and its interaction with the soccer shoe and the ball. Next to a high variability in prices and differences in playability, also the impact on the environment is different in those infill components. Recently organic materials such as cork have been introduced as infill for 3G AT. It has to be highlighted that there is a massive variety in all components of typical 3G AT systems. Developers, manufacturers, and also certifying governing bodies claim that with changes of those components, the playability and relevant characteristics of the pitches might change. Systematic scientific investigations of the effects of the different components of 3G AT systems on player movements and musculoskeletal loading in soccer specific activities are rare, and for some components completely missing.

In order to set quality standards for artificial surfaces to closely mimic the high performance of natural turf (NT) and meet safety requirements, a quality concept has been developed by FIFA. Pitches with the "FIFA quality pro" mark (former 2 star) are specifically tested for a typical usage for professional football up to 20 playing hours per week, while surfaces that are awarded the "FIFA quality" mark (former 1 star) fulfil the specific requirements for recreational, community and municipal football, with typically 40–60 playing hours per week (FIFA, 2015). To date, 562 "FIFA quality pro" pitches and 1652 "FIFA quality" surfaces are installed worldwide (FIFA). In terms of terminology, only artificial pitches meeting these requirements are allowed to call themselves football turf, and consequently qualify for use in official matches.

Since 2004, the governing bodies UEFA and FIFA accepted the use of third-generation (3G) AT for official tournaments and high quality games. Quality approved ATs were already used for games in the World Cup (Women's World Cup (2015), U-17 Women and U-20 Women's and Men's World Cup (on several stadium and training fields, e.g.: Peru 2005, Canada, 2009, 2014, 2015, Jordan 2016, Azerbaijan 2017)), its qualifying games as well as Champions League or Europa League games and others (FIFA, 2015), highlighting the growing importance of this surface at an international level. In 2015, the first World Cup game for Women's football was played with 52 matches exclusively on AT. The decision that AT was the playing surface for this event was not without much criticism from players though, which resulted in a lawsuit against FIFA and the Canadian Soccer Association at the Ontario Human Rights Tribunal in Toronto under the basis of gender discrimination (Bucholtz, 2014). Over 50 female players from around the world raised their claim that men would not play a world cup on AT as an "unsafe" surface. Even though the lawsuit was withdrawn by the players in January 2015, it showed that playing on AT still is under much scrutiny by the performers themselves. It has to be highlighted that AT is generally not seen by the governing bodies as a replacement for NT, but as an alternative surface for areas where climatic conditions make it difficult to grow and maintain an adequate natural surface.

Even though artificial playing surfaces enormously developed over the last years with both producers as well as governing bodies making substantial efforts to bring the game on AT as close as possible to the game played on NT (FIFA, 2010), the actual movement and physiological demands that artificial surfaces impose on the athletes are not very well understood (and we only now are starting to get a better impression of its effect). Given the growing importance and the persistent reservations against the artificial surface, it is however surprising how little consideration performance related aspects for AT received in the football research. This chapter aims to give an overview of the existing research, investigating if the use of artificial surfaces instead of natural ones imposes different technical, tactical and performance related demands on the players and alters the player–surface and ball–surface interaction. While the biomechanical analysis of isolated football-specific maneuvers such as sprint, cut or turn on different surfaces provides preliminary data on single aspects of the game and spotlights biomechanical demands under controlled conditions, which might not be possible to evaluate in a real-game situation, this isolated analysis might not be representative of the physical demand and characteristics of an actual football situation and may not elicit fatigue to the same extent as match-play (Potthast *et al.*, 2010). Therefore more recent studies tried to mimic football exercises to generate a better understanding of the physical demand (Nedelec *et al.*, 2013; Tessitore *et al.*, 2012; Stone *et al.*, 2014, Hughes *et al.*, 2013), recovery (Nedelec *et al.*, 2013; Stone *et al.*, 2014) and perceived intensity (Nedelec *et al.*, 2013) of the game. The variety of existing artificial surface types, each exposing different characteristics depending on fiber length, infill material, etc., as well as the natural surfaces characteristics depending on wear and tear, climate and level of maintenance makes it difficult to draw general conclusions however. For example by changing one component (infill) the differences between ATs can become greater than those between a NT and an AT system (Potthast *et al.*, 2010), an aspect which will challenge future research. Additionally, the specific mechanical characteristics and quality certification of the investigated surface are often not detailed stated in the published research and the surface is commonly referred to as 3G AT. Where available the quality and surface characteristics were included in this chapter, otherwise also the general 3G AT acronym was used.

3. Mechanical characteristics

Due to the artificial nature of AT, surface characteristics such as stiffness and shoe–surface interface can be different to NT. The shoe needs to interlock with the playing surface to generate traction, and different surface characteristics might affect the ability to accelerate, decelerate or change direction (Fleming, 2011). As covered in more detail in the previous chapter in 6.1 *Traction: Shoe-surface interface*, the optimal zone of traction will allow for optimal performance and may minimize injury risk (Luo & Stefanyshyn, 2011; Smeets *et al.*, 2012, Nigg & Segesser, 1988), while excessive tractions resistance will cause foot sticking (Shorten *et al.*, 2003) and high levels of rotational traction are identified as contributor to cause injury (Thomson *et al.*, 2015).

Certified artificial surfaces are required to meet regulatory standards regarding shock absorption, vertical deformation, energy restitution and linear and rotational traction (FIFA, 2010), with the certification for acceptance of AT systems being based on purely mechanical tests of surface components (fibers, infill, elastic layer) (Potthast *et al.*, 2010). While the application of mechanical testing is an important procedure for the identification of surface properties (Charalambous *et al.*, 2016), the related scientific literature and the results of more current studies question the representativeness of these test for the human–surface interaction (Charalambous *et al.*, 2016; Potthast *et al.*, 2010; Fleming, 2011; Fleming *et al.*, 2005; Verhelst, 2010) as various research adds to the theory that athletes adapt their leg stiffness to the stiffness of the surface on which they move (Farley *et al.*, 1998; Ferris *et al.*, 1998). For example, in heel–toe running, softer midsoles of running shoes did not lead to lower ground reaction forces

(Nigg *et al.*, 1995; Shorten, 2002; Clarke *et al.*, 1985). Also when running on hard to soft surfaces, the highest peak loading rate was imposed by a surface that was ranked joint-lowest in terms of hardness and the hardest surface led only to the second highest peak loading rate (Stiles *et al.*, 2011). A further study showed that athletes increased their leg stiffness (decreasing ROM) in response to decreasing surface stiffness (Kerdok *et al.*, 2002). Additionally surface stiffness showed no significant effect on sprinting time, ground contact time and step length (Kerdok *et al.*, 2002, Stafilidis & Arampatzis, 2007). Charalambous *et al.* (2016) state that, while measures by the Artificial Athlete imply a high correlation between impact force and surface stiffness, measures of the human–surface ground reaction force have revealed no correlation between the peak impact force and the stiffness of the surface (Nigg & Yeadon, 1987; Nigg, 1990; Dixon *et al.*, 1999). With respect to linear friction tests, Verhelst (2010) found a poor relationship between the results of pure mechanical linear friction test and the perceived grip and shuttle run performance of athletes, while rotational friction measures seemed to be positively related. In this line of discussion, Potthast *et al.* (2010) highlight that the human response cannot be predicted by only analyzing the mechanical surface properties and a deeper understanding of player–surface interaction and how this depends on alterations of surface properties is needed in order to improve or justify existing standards for certification systems of AT.

4. Biomechanical aspects

4.1 Technical skills

Probably one of the most referenced research papers with respect to player perception and performance changes on artificial turf is the study by Andersson *et al.* (2008). The authors analyzed the playing impression of 72 male and 21 female Swedish elite football players, as well as the movement pattern and ball skills of a subgroup. Subjectively the majority of the male players, who were regular grass players, perceived playing on AT physically harder than on NT, and running without the ball, making a precise pass as well as ball control and taking a shot to be more difficult on AT. They were neutral about the feel of running with the ball and the speed of passing the ball. Male players who regularly play on AT also show a negative impression of AT and reported that playing on NT was easier both physically and technically. The female players stated a neutral impression with respect to AT, feeling that surface did not influence the game in general, with no difference in the physical strain of the game, or the difficulty in making a precise pass, controlling the ball, or taking a shot. They reported however, that it was easier to run with the ball and to pass the ball when playing on AT. The perception of the male players is in discrepancy to analyzed movement parameters, which show generally no difference between AT and NT in terms of total distance covered, high intensity running, numbers of sprints, standing tackles or headers per game. However, about 50% fewer sliding tackles and 20%

more short passes mainly appearing in the midfield-to-midfield-zone were observed on AT. The reduced number of sliding tackles could be explained by the perceived risk of burns and other short-term injuries, resulting in a less aggressive play on AT. On the other hand, the AT possibly allowed an improved ball control as less unsuccessful short passes occurred on AT than on NT, which would make possession play or "power play" much easier. This study was the first one to demonstrate that subjective and objective measures of the movement on AT do not necessarily coincide, and need further evaluation. Similar results were observed by a FIFA study (FIFA, 2008), which analyzed 110 elite competitive matches. No differences between 3G AT and NT surface on the length, direction and accuracy of passes, the effective playing time, the number of on-the-ball activities and possession transitions were shown, but trends for a higher incidence of shots, goals and passes on 3G AT were observed.

These two studies are the only studies investigating the tactical play on artificial surfaces compared to natural grass surfaces and suggest little differences with possibly more power play, and less tackle activities on AT. In comparison to playing football on AT, asphalt and beach sand, Brito *et al.* (2012) noticed the number of high intensity runs to be higher on asphalt than on AT and similar on sand. The perceived exertion, however, was lower on asphalt than on AT and sand. Tessitore *et al.* (2012) investigated heart rate responses and technical aspects of play between a clay surface and a 3G AT. In a setting of 5-a-side youth soccer matches, 22 young male soccer players (age 8.3 ± 0.4 years) were engaged in two consecutive 15 minute periods on each surface. No difference emerged between both pitch surfaces and match periods for heart rate response and match analysis indicators such as type of action, the number of players involved in an action, the number of passes performed in a collective action, precision of the shots, lost balls, ball interceptions, dribbles and tackles. These two studies indicate that playing on asphalt and sand lead to small performance changes, while in youth football, main performance was unaffected by pitch surface.

4.2 Running and sprinting

During a game, elite outfield players typically cover distances of 10 to 12 km on NT (Andersson *et al.*, 2008). While mainly dominated by low intensity activities, they perform up to 250 brief high intensity actions during a match and produce blood lactate concentrations of 10–14mmol/L (Bradley *et al.*, 2009; Krustrup *et al.*, 2006). Sprinting occurs every 90–120s and high intensity efforts occur every 40–70s (Bradley *et al.*, 2010; Reilly & Thomas, 1976; Reilly, 1994).

Meijer *et al.* (2006) were the first to investigate, with a pilot study of four participants, the ground reaction force of running as football specific movement on 1 star and 2 star 3G AT under standardized conditions with preferred jogging speed, a run at 17.5 km/h and a full sprint. They concluded that surface characteristics influence the loading of the human musculoskeletal system more subtly (force oscillations, magnitude of vertical force, contact time and push-off forces, time of force-peaks) than initially anticipated.

In his doctoral thesis, Verhelst (2010) investigates the influence of different surfaces on perception and performance expressed as total running time in a shuttle run, also requiring multiple 180° turns. Twelve male soccer players of medium to high level performed the testing maneuver on a slightly soggy NT and two types of 3G AT (AT1, AT2), both fulfilling 1 or 2 star FIFA requirements for rotational and translational friction. The average total running time was significantly higher (ca 3%) on NT than on AT1, but similar on AT2. The slip was quantified by the players with a visual-analog-scale (VAS – ratings from 0–10) and was perceived highest on NT, which corresponded with the longest running time.

Potthast *et al.* (2010) refer to Verhelst *et al.* (2007) and Verhelst (2010), who set the 2D kinematics and kinetics of a standing start at two types of 3G AT in focus. The movement of seven male recreational soccer players was analyzed on a surface with standard styrene-butadiene rubber (SBR) infill, and a thermoplastic elastomer (TEP) infill. Significant differences were observed between the two surfaces, with the SBR surface showing a higher maximum coefficient of friction, horizontal impulse, contact time and knee flexion angle at toe-off. The latter three suggest a higher grip on SBR as the leg can be extended more without the risk that the foot starts to slip. This result demonstrates that changes in just one turf component can substantially change performance and hence interpretation of how "the AT" surfaces affect performance and perception needs to be made with caution.

Another study from national collegiate American football (Gains et al., 2010) splits the performance components in straight line running (40yd dash sprint) and a pro-agility test, requiring multiple change-of-direction maneuvers. Results indicate that no significant differences between AT and NT in 40yd dash times exist, but that pro-agility times were significantly faster on AT than on NT (4.49 ± 0.28s, vs 4.64 ± 0.33s). In summary these studies show 1) subtle changes in GRF due to surface type, 2) slower shuttle run times on NT with lower grip on NT, and 3) no differences in straight line running but slower performance timings in the pro-agility test requiring change of direction tasks on NT. It appears that surface differences might become more pronounced in tasks demanding rotational traction, such as a change-of-direction task than straight line running.

4.3 Cutting

Throughout matches, players execute a range of movement skills and skills involving up to 180 activities "on the ball" (Andersson *et al.*, 2008; Bloomfield *et al.*, 2007) – the average time in ball contact for a player is about 2% of the entire game though (Bangsbo, 1994) – and over 1250 changes in movement type (Andersson *et al.*, 2008, Bloomfield *et al.*, 2007). While the most important movement skill is repeated sprint ability, due to the intermittent nature of soccer (Chaouachi *et al.*, 2010), also frequent accelerations and directional changes occur (Bloomfield *et al.*, 2007, Carling *et al.*, 2008). Two studies were identified investigating the kinematics of a cutting maneuver in a male (Lake & Underdown,

2011) and female (Strutzenberger *et al.*, 2014) population. Lake and Underdown (2011) analyzed the V-cut maneuver (135° change-of-direction task) of 19 semi-professional male players on six NT and three 3G AT pitches in three different countries. Average ground contact time and exit velocity as well as peak foot displacements, velocities and accelerations during the initial and final stages showed little differences across the various surfaces. The authors suggested that there was more loading demand placed on the ankle joint on AT as inversion motion of the ankle during the turn appeared to be similar on the natural surfaces but the rate and range of inversion was significantly increased on the AT surfaces. Additionally, the development of traction appeared to be more rapid on the AT surfaces, while peak traction was similar across all the tested surfaces. Lake and Underdown (2011) highlight a high variability in the outdoor data and question whether some of the chosen parameters were the right ones to discriminate between surfaces. In a female population, Strutzenberger *et al.* (2014) investigated the kinematics in the early deceleration phase of an unanticipated cutting (30° and 60°) task of eight university-level female football players on an outdoor 3G AT (FIFA 2 star) and NT. With the same run-up velocity (4–5m/s) the female players displayed a trend to a reduced valgus and internal rotation at the knee on the AT. The ankle joint during the first 20% of the cutting stance phase demonstrated a trend of the surface for increased dorsiflexion and inversion on AT. The authors conclude that the kinematics displayed by the female players do not indicate an increased risk for ACL injury, but that the alterations of ankle kinematics need to be further investigated.

4.4 Kicking

As player perception and the plantar loading pattern of the supporting foot indicate that soccer players interact differently with the different types of surface systems, Potthast (2010) aimed for a deeper understanding of the influence of different infill materials (sand and rubber vs. rubber granulate only) on the goal kick in reference to a NT. On the AT with a sand and rubber infill, the players showed the same run-up behavior but interrupted the forward movement of the center of gravity more cautiously than on the other surfaces. This coincided with a slightly more backward orientation of the supporting leg on sand and rubber, while the orientation of the supporting leg on the rubber infill AT and NT was similar. It was interpreted that this alteration meant that players adopted a more cautious strategy on the sand and rubber infill surface and resulted in the lowest ball speed and kicking accuracy compared with the other two surfaces. In terms of kicking performance, the possibility of transferring the impulse of the center of gravity into the angular impulse of the lower extremity is essential, and a reduced deceleration might be detrimental to the ability to generate a hard kick. This study adds further evidence to the observation that the interaction between player and surface strongly depends on the characteristics of the surface.

The used methods in these reported studies are mainly performance measures in terms of sprint times and kinematic measures for example. The actual loading on the musculoskeletal system via the linkage of kinematic, kinetic and electromyographic data has not been quantified and remains speculative. This is given due to the principal problem of analyzing the effects of NT and AT on the musculoskeletal load in football specific situations. Simulating surface conditions in a laboratory environment including force measurement devices needed for kinetic data or installing laboratory equipment at NT and AT pitches creates technical challenges, especially regarding the adequate placement of the surface on force plates to determine ground reaction forces with high quality.

4.5 Temperature

The effect of the temperature on the artificial playing surface might take effect, when temperatures come close to zero and harden the surface (Charalambous *et al.*, 2016), or in hot weather conditions, where the playability of synthetic turf may be hampered, since the rubber and other materials in AT absorb heat at a rapid rate (Thoms *et al.*, 2014; Petrass *et al.*, 2014). This leads to surface temperatures that are higher than surrounding air temperatures and temperatures of other surfaces (Simon, 2010; Dragoo & Braun, 2010; Williams & Pulley, 2003). For example at the Women's World Cup 2015 in Canada, the AT surface reached 49° C for the opening match of the tournament with air temperature recorded at 23° C (Bossi, 2015). Athletes addressed one major aspect that occurred with the higher temperatures, as they point out, these temperatures leave blisters and cuts on the player's feet – a fact also observed by Williams and Pulley (2003) – and it is difficult to get a good grip in the shoe due to increased sweating (Bossi, 2015). Further concerns related to the heated-up surfaces are, a) a possible increased internal stress for the players as the temperature could increase dehydration and fatigue (Simon, 2010), b) a possible increased risk due to higher off-gazing of hazardous elements from infill materials (EHHI, 2010), and c) a change in rotational surface stiffness with varying temperatures possibly influencing the risk of injury (Torg *et al.*, 1996).

It is surprising, that even though the player response to the surface temperature was addressed by some researchers as factor needing further investigation (e.g. Potthast *et al.*, 2010), heat aspects are still underrepresented in scientific literature and available information was mainly found by the authors in press releases (Bossi, 2015) and research reports (Simon, 2010, Williams & Pulley, 2003). For example Williams and Pulley (2002) tested the temperatures of the University of Brigham Young University's American football AT, of the football (soccer) AT and of the NT. On the tested days with average air temperature of 27°C, average surface temperatures for the artificial surfaces were approx. 47°C while NT had 26°C. The authors notice, when surface temperature reaches 50°C it takes less than 10 min to cause injury to skin. Irrigation of the synthetic turf had a significant result in cooling but lasted only twenty minutes. However the authors do not provide detailed data of the surface used (manufacturer, FIFA quality approval), making it difficult to rank the tested surface within the variety of surfaces available. The understanding of the heat absorption of the current surfaces in order to be able to develop cooler surfaces in the future is currently a research topic (e.g. Devitt *et al.*, 2007, McNittt *et al.*, 2008, Petrass *et al.*, 2014).

While the concerns about excessive heat for player well-being and performance have not yet been properly addressed in the scientific research, Charalambous *et al.* (2016) laid a first puzzle piece in cooler conditions and investigated the mechanical properties of two 3G AT surface and the kinematics of hurdle jump to sprint acceleration movement on a cold ($1.8°$–$2.4°C$) and warm (14.5–$15.2°C$) 3G AT surface performed by four amateur soccer players. Mechanical testing showed that the temperature difference had a significant influence on the mechanical properties of the AT, including force absorption, energy restitution, rotational resistance and the head injury criterion. In terms of a player's movement, no differences were found for the landing, but both step length and contact time of the initial step after the landing were significantly longer on the warm surface. In addition, a significant range of motion and joint angular velocity differences were found. However, no conclusion can be drawn if these responses of the athletes are due to a colder air and surface temperature, and if these differences would have appeared when playing on NT as well. In summary, it is known that surface temperature are increased on AT, but the actual player response remains a field for future research. It is generally stated, that the unplayability of 3G AT due to too high temperature does not outweigh the degree to which natural fields are compromised during rain and snow (Simon, 2010).

4.6 Physiological responses

The declines demonstrated in physiological responses toward the end of 90-minute elite soccer matches are attributed to an increase in thermal stress and a decrease to fatigue (Bangsbo, 1994; Tessitore *et al.*, 2012). As artificial surfaces can provoke differences in isolated movement alterations, these changes might also contribute to altered physiological responses and fatigue effects imposed by different surface properties (Hughes *et al.*, 2013).

4.7 Running

Sassi *et al.* (2011) evaluated the metabolic cost of running (Cr) on 3G AT, NT and on asphalt track of eight amateur male football players. Each player completed a six-minute run on each surface with three running speeds (2.22, 2.78, 3.33 m/s). Athletes displayed similar Cr values for both AT and NT (~ 4.20 J/kg/m), both being significantly higher by 5% compared to the asphalt track. With this result, authors don't confirm the subjective opinion of athletes, who report that performance on AT felt more strenuous than on NT (Andersson *et al.*, 2008, Poulos *et al.*, 2014). They also are contrary to findings from Di Michele *et al.* (2009), who found higher blood-lactate accumulation and heart rate when running with the same speed on AT compared to NT. Sassi *et al.* (2011) however criticize the latter paper, by saying that no data on surface characteristics were provided, hence it is difficult to attribute this change to the artificial surface.

4.8 Match play and simulated soccer protocols

A criticism of the analysis of individual soccer movements is that the analysis might not reflect the metabolic state athletes experience over a football match. On the one hand, a decline in skilled performance can be seen when the player is fatigued (Russell *et al.*, 2011, Stone & Oliver, 2009), and on the other hand fatigue is associated with increased risk of injury (Greig & Siegler, 2009, Woods *et al.*, 2004). If playing on AT imposes a higher physical demand for the player than playing on a grass surface, these effects might be more pronounced over the course of a game. Therefore, it seems crucial to investigate the player performance in a setting as closely as possible related to football-specific demands (Hughes *et al.*, 2013, Stone *et al.*, 2014, Nedelec *et al.*, 2013, Tessitore *et al.*, 2012).

Nedelec *et al.* (2013) investigated the influence of surface type on physical performance and recovery. Thirteen male professional football players, who were familiar with playing on AT underwent a 90 minute football-specific exercise (SAFT90 (Small *et al.*, 2010)). Before and at three timed (immediately, 24h and 48h) points after the 90 minute football-specific exercise performance tests (Squat Jump (SJ), Counter-Momement-Jump (CMJ), 6-s sprint, isometric hamstring torque) and quality of sleep, fatigue, muscle soreness, stress and total quality recovery were analyzed and compared. Mainly trivial to small effects were shown, with no differences in sprint performance, mean heart rate, fluid loss and rating on intensity. Diverging results were found for the jump test, where the SJ was closer to the baseline values on NT than on AT 48 hours after the exercise, while the CMJ displayed a faster recovery on AT than on NT. No surface–time interaction effects were shown for quality of sleep, muscle soreness for 14 out of the 17 investigated muscles of the hip and lower extremity, stress and total recovery with only trivial to small differences. However soreness in quadriceps immediately after the 90 min field test, in gluteus 24 hours after the test and in hamstring 48 hours after were all reported to be moderately lower (31–46%) on NT than on AT. Overall the authors concluded that there is no evidence to indicate that exercise results in greater fatigue and delayed recovery process. They notice, though, that their field test did not involve contact situations such as tackles and collisions which might have affected the overall fatigue of the players.

Hughes *et al.* (2013) investigated on high quality 3G AT and NT with average air temperature of 27.5°C physiological and performance parameters in a soccer simulation protocol (SSP) (Stone *et al.*, 2011), which also includes high-speed turns and intermittent periods of repeated sprints. Immediately before and after each simulation trial, 17 semi-professional outfield football players performed tests of agility, sprint speed and vertical jump height. None of these tests displayed interaction effects between surface and time, indicating that playing surface did not have an additional effect of the development of fatigue. The turns within the sprint agility-run were generally performed more quickly on NT than on AT, while the cuts were similar. The authors conclude that physiological responses to simulated soccer activity do not differ between the high-quality NT and AT used. The performance tests, however, show that playing on AT lead to small alterations especially where changes of direction are required. This is in agreement with previous research from isolated movements stating that turning movements might elicit different biomechanical responses on AT compared to NT (Villwock *et al.*, 2009, Gains *et al.*, 2010, Verhelst, 2010, Verhelst *et al.*, 2007).

Stone *et al.* (2014) used the same SSP with the aim of investigating within-match responses and post-match recovery in the 48 hours following the 90 minute simulated football activity between NT and 3G AT (1 star) in players who do not regularly play or train on AT surfaces, and might therefore be more sensitive to an acute shift between playing surfaces. Eight male Welsh Division 1 soccer players performed immediately prior, as well as immediately after, 24 hours after and 48 hours after each SSP measures of sprint, jump and agility performance. Similar to Nedelec *et al.* (2013) and Hughes *et al.* (2013) no differences in physiological parameters were found during the SSP between the two surfaces. During the simulation protocol, repeated sprint times were slower on NT only, while single 15m sprint and agility times showed no differences. None of the performance variables assessed before and immediately after the soccer simulation protocol shows an interaction effect between surface and time, and only the L-agility run revealed faster times on the NT. Interestingly no difference between playing surfaces in the recovery of performance variables, Creatine Kinase, or perception of muscle soreness in the 48 hours following the simulation protocol was observed.

Overall these studies show that while only a few of performance measures showed a significant effect (repeated sprint (Stone *et al.*, 2014), agility (Hughes *et al.*, 2013)) over time, the different surfaces did not lead to a different fatigue response in immediate performance and physiological measures. Also for the recovery of the subsequent following two days, Stone *et al.* (2014) did not find in either a subjective and objective measure of muscle damage a difference, while Nedelec *et al.* (2013) reported only little changes due to surface type. In only three out of 17 investigated muscles small effects of lower muscle strain were observed when players performed on NT, while higher decrements in squat jump height but lower decrements in hamstring peak torque occurred following the 48 hours after match play on AT. It is questionable whether these small differences, are likely to significantly alter the overall response to match play.

5. Conclusion

This chapter has examined current research driven evidence relating to the influence of the introduction of AT on player performance. It is surprising how little consideration performance related aspects for the play on AT is given in the football-specific research. This might be partly originated in the methodological difficulties of conducting ecological valid research including kinematic, kinetic and muscular activity with respect to surface type. The section of this chapter on mechanical surface characteristics and performance also emphasizes the need for ecological variety within testing procedures to ensure the mechanical tests more closely reflecting the player specific movements. Secondly biomechanical aspects of player performance were reviewed and it is suggested that from a performance perspective, fewer tackles and more short passes occur on AT. In movements that require changes of direction and lower limb rotation artificial turf has a greater influence on performance. In addition it was observed that changes of one turf component can substantially change overall performance and consequently the explanation of how AT surfaces effect performance and perception needs to be made with caution. One key aspect that can be concluded from this section is the need for further investigation into the biomechanical characteristic of performance to understand and explain more fully the changes in technique due to the AT, which needs to be player specific and individually driven. This chapter has highlighted that the environmental factor of surface temperature has been referred to as problematic. However, in this area there is a lack of research which has specifically looked at the influence of temperature on player performance and whether changes in technique occur. Finally physiological responses have demonstrated that there is similar physiological response of players between AT and NT. However, in real match situations, physiological differences due to difference in match play conditions cannot yet be excluded. This chapter has highlighted the need to examine the influence of AT on performance in more detail than currently exists. A clear focus on the underlying biomechanics of the skills and techniques is required from a playing-position and player specific perspective. Analysis of movement patterns, joint work, and coordination are future directions that need interrogating as AT becomes embedded across the sport.

References

Andersson, H., Ekblom, B. & Krustrup, P. (2008). Elite football on artificial turf versus natural grass: movement patterns, technical standards, and player impressions. *Journal of Sports Sciences*, 26, 113–122.

Bangsbo, J. (1994). The physiology of soccer – with special reference to intense intermittent exercise. *Acta Physiologica Scandinavica*, 619, 1–155.

Bloomfield, J., Polman, R. & O'Donoghue, P. (2007). Physical demands of different positions in FA Premier League Soccer. *Journal of Sports Science and Medicine*, 6, 63–70.

Bossi, D. (2015). Women's World Cup: Matildas say synthetic turf 'like hot coals'. *Sydney Morning Herald*.

Bradley, P. S., Di Mascio, M., Peart, D., Olsen, P. & Sheldon, B. (2010). High-intensity activity profiles of elite soccer players at different performance levels. *Journal of Strength and Conditioning Research*, 24, 2343–2351.

Bradley, P. S., Sheldon, W., Wooster, B., Olsen, P., Boanas, P. & Krustrup, P. (2009). High-intensity running in English FA Premier League soccer matches. *Journal of Sports Sciences*, 27, 159–168.

Brito, J., Krustrup, P. & Rebelo, A. (2012). The influence of the playing surface on the exercise intensity of small-sided recreational soccer games. *Human Movement Science*, 31, 946–956.

Bucholtz, A. (2014). American-led group of top women's soccer players files suit against Canada's plan to hold 2015 Women's World Cup on turf, but will it amount to anything more than the ski jumping law-suit? Online Blog. http://sports.yahoo.com/blogs/eh-game/american-led-group-of-top-women-s-soccer-players-files-lawsuit-against-canada-s-plan-to-hold-women-s-world-cup-on-turf–but-will-it-amount-to-anything-more-than-the-women-s-ski-jumping-lawsuit-223643909.html.

Carling, C., Bloomfield, J., Nelsen, L. & Reilly, T. (2008). The role of motion analysis in elite soccer: contemporary performance measurement techniques and work rate data. *Sports Medicine*, 38, 839–862.

Chaouachi, A., Manzi, V., Wong Del, P., Chaalali, A., Laurencelle, L., Chamari, K. & Castagna, C. (2010). Intermittent endurance and repeated sprint ability in soccer players. *Journal of Strength and Conditioning Research*, 24, 2663–2669.

Charalambous, L., Von Lieres und Wilkau, H., Potthast, W. & Irwin, G. (2016). The effects of artificial surface temperature on mechanical properties and player kinematics during landing and acceleration. *Journal of Sport and Health Science*, 5, 355–360.

Clarke, T. E., Cooper, L. B., Hamill, C. L. & Clark, D. E. (1985). The effect of varied stride rate upon shank deceleration in running. *Journal of Sports Sciences*, 3, 41–9.

Devitt, D. A., Young, M. A., Baghzouz, M. & Bird, B. M. (2007). Surface temperature, heat loading and spectral reflectance of artificial turfgrass. *Journal of Turfgrass and Sport Surface Science*, 83, 68–82.

Di Michele, R., Di Renzo, A. M., Ammazzalorso, S. & Merni, F. (2009). Comparison of physiological responses to an incremental running test on treadmill, natural grass, and synthetic turf in young soccer players. *Journal of Strength and Conditioning Research*, 23, 939–945.

Dragoo, J. L. & Braun, H. J. (2010). The effect of playing surface on injury rate: A review of the current literature. *Sports Medicine*, 40, 981–990.

Fédération Internationale de Football Association. FIFA recommended pitches worldwide. Online database. http://quality.fifa.com/en/Football-Turf/FIFA-recommended-pitches-wordwide/#/index.

Fédération Internationale de Football Association (2008). Does the game change on football turf? *FIFA Turf Roots Magazine*, 3, 31–34.

Fédération Internationale de Football Association (2010). FIFA Quality concept for football turf. http://www.fifa.com/mm/document/afdeveloping/pitchequip/fqc_football_turf_folder_342.pdf.

Fédération Internationale de Football Association (2015). Member Association Handbook FIFA Quality Programme for Football Turf. October 2015. http://quality.fifa.com/globalassets/ma-handbook-football-turf-en.pdf.

Fleming, P. (2011). Artificial turf systems for sport surfaces: Current knowledge and research needs. *Journal of Sports Engineering and Technology*, 225, 43–63.

Fleming, P. R., Young, C., Roberts, J. R., Jones, R. L. & Dixon, N. (2005). Human perceptions of artificial surfaces for field hockey. *Sports Engineering*, 8, 121–136.

Gains, G. L., Swedenhjelm, A. N., Mayhew, J. L., Bird, H. M. & Houser, J. J. (2010). Comparison of speed and agility performance of college football players on field turf and natural grass. *Journal of Strength and Conditioning Research*, 24, 2613–2617.

Greig, M. & Siegler, J. C. (2009). Soccer-specific fatigue and eccentric hamstrings muscle strength. *Journal of Athletic Training*, 44, 180–184.

Hughes, M. G., Birdsey, L., Meyers, R., Newcombe, D., Oliver, J. L., Smith, P.M., Stembridge, M., Stone, K. & Kerwin, D. G. (2013). Effects of playing surface on physiological responses and performance variables in a controlled football simulation. *Journal of Sports Sciences*, 31, 878–886.

Kerdok, A. E., Biewener, A.A., Mcmahon, T. A., Weyand, P. G. & Herr, H. M. (2002). Energetics and mechanics of human running on surfaces of different stiffnesses. *Journal of Applied Physiology*, 92, 469–478.

Krustrup, P., Mohr, M., Steensberg, A., Bencke, J., Kjaer, M. & Bangsbo, J. (2006). Muscle and blood metabolites during a soccer game: Implications for sprint performance. *Medicine and Science in Sports and Exercise*, 38, 1–10.

Lake, M. & Underdown, T. (2011). Traction and lower limb motion during rapid turning on different football turf surfaces. *Footwear Science*, 3: suppl, S92–S93.

Luo, G. & Stefanyshyn, D. (2011). Identification of critical traction values for maximum athletic performance. *Footwear Science*, 33, 127–138.

McNittt, A., Petrunak, D. M. & Serensits, T. J. (2008). Temperature amelioration of synthetic turf surfaces through irrigation. In: *Proceedings of 2nd International Conference on Turfgrass*, pp. 573–581.

Meijer, K., Detherms, J., Savelber, H. & Williams, P. (2006). The influence of third generation artificial soccer turf characteristics on ground reaction forces during running. In H. Schwameder, G. Strutzenberger, V. Fastenbauer, S. Lindinger & E. Müller (Eds.), Proceedings of *24th Symposium on Biomechanics in Sports*, Salzburg, Austria: University of Salzburg, pp. 1–4.

Nedelec, M., Mccall, A., Carling, C., Le Gall, F., Berthoin, S. & Dupont, G. (2013). Physical performance and subjective ratings after a soccer-specific exercise simulation: comparison of natural grass versus artificial turf. *Journal of Sports Sciences*, 31, 529–536.

Nigg, B. M., Cole, G. K. & Brueggemann, G. P. (1995). Impact forces during heel–toe running. *Journal of Applied Biomechanics,* 11, 407–432.

Nigg, B. M. & Segesser, B. (1988). The influence of playing surfaces on the load on the locomotor system and on football and tennis injuries. *Sports Medicine*, 5, 375–385.

Petrass, L. A., Twomey, A. & Harvey, J. T. (2014). Understanding how the components of a synthetic turf system contribute to increased surface temperature. *Procedia Engineering*, 72, 943–948.

Potthast, W. (2010). Motion differences in goal kicking on natural and artificial soccer turf systems. *Footwear Science*, 2, 29–35.

Potthast, W., Verhelst, R., Hughes, M., Stone, K. & De Clercq, D. (2010). Football-specific evaluation of player–surface interaction on different football turf systems. *Sports Technology*, 3, 5–12.

Poulos, C. C., Gallucci, J. Jr., Gage, W. H., Baker, J., Buitrago, S. & Macpherson, A. K. (2014). The perceptions of professional soccer players on the risk of injury from competition and training on natural grass and 3rd generation artificial turf. *BMC Sports Science, Medicine and Rehabilitation*, 6, 11.

Reilly, T. (1994). Physiological profile of the player. In B. Ekblom (Ed.), *Handbook of Sports Medicine and Science, Football (soccer)* (pp. 78–94), Oxford: Blackwell Scientific Publications.

Reilly, T. & Thomas, V. (1976). A motion analysis of work-rate in different positional roles in professional football match play. *Journal of Human Movement Studies*, 2, 87–89.

Russell, M., Benton, D. & Kingsley, M. (2011). The effects of fatigue on soccer skills performed during a soccer match simulation. *International Journal of Sports Physiology and Performance*, 6, 221–233.

Sassi, A., Stefanescu, A., Menaspa, P., Bosio, A., Riggio, M. & Rampinini, E. (2011). The cost of running on natural grass and artificial turf surfaces. *Journal of Strength and Conditioning Research*, 25, 606–611.

Shorten, M. (2002). The myth of running shoe cushioning. In S. Ujihashi & S. J. Haake (Eds.), *The IV International Conference on the Engineering of Sport*, pp.1–6, Kyoto, Japan.

Shorten, M., Hudson, B. & Himmeslbach, J. (2003). Shoe–surface traction of conventional and in-filled synthetic turf football surfaces. In Milburn et al., (Ed.) *Proceedings of 15th International Congress of Biomechanics*, (p. 6), Dunedin, New Zealand. University of Otaga.

Simon, R. (2010). Review of the impacts of crumb and rubber in artificial turf applications. *Green Manufacturing and Sustainable Manufacturing Partnership*. Berkeley, CA: University of California Laboratory for Manufacturing and Sustainability.

Small, K., Mcnaughton, L., Greig, M. & Lovell, R. (2010). The effects of multidirectional soccer-specific fatigue on markers of hamstring injury risk. *Journal of Science and Medicine in Sport*, 13, 120–125.

Smeets, K., Jacobs, P., Hertogs, R., Luyckx, J. P., Innocenti, B., Corten, K., Ekstrand, J. & Bellemans, J. (2012). Torsional injuries of the lower limb: An analysis of the frictional torque between different types of football turf and the shoe outsole. *British Journal of Sports Medicine*, 46, 1078–1083.

Stafilidis, S. & Arampatzis, A. (2007). Muscle–tendon unit mechanical and morphological properties and sprint performance. *Journal of Sports Sciences*, 25, 1035–1046.

Stiles, V. H., Guisasola, I. N., James, I. T. & Dixon, S. J. (2011). Biomechanical response to changes in natural turf during running and turning. *Journal of Applied Biomechanics*, 27, 54–63.

Stone, K. J., Hughes, M. G., Stembridge, M. R., Meyers, R. W., Newcombe, D. J. & Oliver, J. L. (2014). The influence of playing surface on physiological and performance responses during and after soccer simulation. *European Journal of Sports Science*, 1–8.

Stone, K. J. & Oliver, J. L. (2009). The effect of 45 minutes of soccer-specific exercise on the performance of soccer skills. *International Journal of Sports Physiology and Performance*, 4, 163–175.

Stone, K. J., Oliver, J. L., Hughes, M. G., Stembridge, M. R., Newcombe, D. J. & Meyers, R.W. (2011). Development of a soccer simulation protocol to include repeated sprints and agility. *International Journal of Sports Physiology and Performance*, 6, 427–431.

Strutzenberger, G., Cao, H. M., Koussev, J., Potthast, W. & Irwin, G. (2014). Effect of turf on the cutting movement of female football players. *Journal of Sport and Health Science*, 3, 314–319.

Tessitore, A., Perroni, F., Meeusen, R., Cortis, C., Lupo, C. & Capranica, L. (2012). Heart rate responses and technical-tactical aspects of official 5-a-side youth soccer matches played on clay and artificial turf. *Journal of Strength and Conditioning Research*, 26, 106–112.

Thoms, A. W., Brosnan, J. T., Zidek, J. M. & Sorochan, J. C. (2014). Models for predicting surface temperatures on synthetic turf playing surfaces. *Procedia Engineering*, 72, 895–900.

Thomson, A., Whiteley, R. & Bleakley, C. (2015). Higher shoe–surface interaction is associated with doubling of lower extremity injury risk in football codes: A systematic review and meta-analysis. *British Journal of Sports Medicine*, 49, 1245–1252.

Torg, J. S., Stilwell, G. & Rogers, K. (1996). The effect of ambient temperature on the shoe–surface interface release coefficient. *The American Journal of Sports Medicine*, 24, 79–82.

Verhelst, R. (2010). Study of ball–surface and player–surface interaction on artificial turf. *PhD Doctoral dissertation*, Ghent University.

Verhelst, R., Verleysen, P., Degrieck, J. & De Clercq, D. (2007). Recent findings of transmitted forces for early accelerations in football. In *proceedings of Workshop "New techniques and technologies to apply in sport studies"*, Valencia, Spain.

Villwock, M. R., Meyer, E. G., Powell, J. W., Fouty, A. J. & Haut, R. C. (2009). Football playing surface and shoe design affect rotational traction. *The American Journal of Sports Medicine*, 37, 518–525.

Williams, C. F. & Pulley, G. E. (2003). *Synthetic surface heat studies*. Brigham: Brigham Young University. http://aces.nmsu.edu/programs/turf/documents/brigham-young-study.pdf.

Woods, C., Hawkins, R. D., Maltby, S., Hulse, M., Thomas, A. & Hodson, A. (2004). The Football Association Medical Research Programme: an audit of injuries in professional football-analysis of hamstring injuries. *British Journal of Sports Medicine*, 38, 36–41.

Epilogue
Future research directions

At the end of this volume, it is worth acknowledging possible future research directions, which would be of help to further expand our understanding of *Football Biomechanics* in the future.

First of all, the limitations of conventional biomechanical research design, currently used in most studies, should be noted. In an experiment, researchers tend to pick a certain number of successful trials while excluding failed trials from their data sets. This illustrates an ideal aspect of techniques, yet ignores the part of missed or half missed shots, which players frequently execute during a football match. Thus, further studies looking at miss-hit or half miss-hit shots would be warranted in the future.

Football kicking motion includes a sudden change of frequency component caused by ball impact. That makes it difficult to adequately remove noise from the measured signal. To date, not many studies acknowledged this type of issue, thereby concealing the true nature of the kicking motion. A time-frequency filter is a best feasible solution. However, this type of filtering is not yet popular in the field of football biomechanics. More applications of this advanced technology to the kicking motion should be emphasized.

Football kicking motion has been, and possibly will be the main focus of football biomechanics research. However, other football actions like ball stopping, goalkeeping skills, heading and ball throwing have gained little attention from researchers. Also, most kicking studies have just focused on maximal effort kicking, in contrast to the frequent situation in which players are required to kick the ball with submaximal effort and/or more accurate targeting. Examining these areas would contribute to a better understanding of football techniques.

Footwear is the main equipment in football. In contrast to the rapidly increasing number of female footballers, there is a dearth of shoe-related research for female players. Women have a different foot shape, and different traction demands as consequence of their different playing style. Because traction on the field is suspected to be strongly linked to knee injuries, research should focus on this aspect. This may help to reduce the high incidence of anterior cruciate ligament injuries in female soccer players. Also, footwear design can influence ball touch, kicking accuracy and injury protection. These aspects (in contrast to the advertising of their products) have largely been ignored by manufacturers. More work on these footwear aspects is needed.

Artificial turf is most likely becoming a new, approved football surface. However, players and coaches still have negative opinions about these new surfaces. Recently, several epidemiological studies were performed to identify the characteristics of artificial turf and their influence on injury and performance perspectives. However, epidemiological evidence does not yet exist sufficiently, and even less biomechanical research has studied the interaction of players with artificial turf surfaces. This kind of research would offer a good opportunity to solve the concerns of coaches and players about playing on artificial turf.

Finally, as highlighted by a number of chapters in this book, it is imperative that we can translate the findings of our scientific studies into coach and athlete friendly interventions. This presents a real challenge to our future work. Frameworks have been outlined in this book to aid design of suitable intervention techniques that are required to alter a player's technique for improved performance, or injury prevention. Much further work is needed here so that we can aid coaches in administering the technical changes that our biomechanical studies have uncovered, and indeed make sure these alterations remain after a period of time.

Index